I0883220

Public Engagement on Genetically Modified Organisms

WHEN SCIENCE AND CITIZENS CONNECT

WORKSHOP SUMMARY

Holly Rhodes and Keegan Sawyer, Rapporteurs

Roundtable on Public Interfaces of the Life Sciences

Board on Life Sciences

Division on Earth and Life Studies

Board on Science Education

Division of Behavioral and Social Sciences and Education

NATIONAL RESEARCH COUNCIL
OF THE NATIONAL ACADEMIES

THE NATIONAL ACADEMIES PRESS
Washington, D.C.
www.nap.edu

THE NATIONAL ACADEMIES PRESS 500 Fifth Street NW Washington, DC 20001

NOTICE: The project that is the subject of this report was approved by the Governing Board of the National Research Council, whose members are drawn from the councils of the National Academy of Sciences, the National Academy of Engineering, and the Institute of Medicine. The members of the committee responsible for the report were chosen for their special competences and with regard for appropriate balance.

The Roundtable on Public Interfaces of the Life Sciences is funded by the Gordon and Betty Moore Foundation through Grant GBMF3869, the Burroughs Wellcome Fund through Grant 10002067, the Howard Hughes Medical Institute through Grant 10001304, Monsanto through Grant 10001447, DuPont, and the National Academy of Sciences.

Any opinions, findings, conclusions, or recommendations expressed in this publication are those of the author(s) and do not necessarily reflect the views of the organizations or agencies that provided support for the project.

International Standard Book Number-13: 978-0-309-37421-7
International Standard Book Number-10: 0-309-37421-9

Additional copies of this report are available for sale from the National Academies Press, 500 Fifth Street, NW, Keck 360, Washington, DC 20001; (800) 624-6242 or (202) 334-3313; Internet, http://www.nap.edu.

Copyright 2015 by the National Academy of Sciences. All rights reserved.

Printed in the United States of America.

THE NATIONAL ACADEMIES
Advisers to the Nation on Science, Engineering, and Medicine

The **National Academy of Sciences** is a private, nonprofit, self-perpetuating society of distinguished scholars engaged in scientific and engineering research, dedicated to the furtherance of science and technology and to their use for the general welfare. Upon the authority of the charter granted to it by the Congress in 1863, the Academy has a mandate that requires it to advise the federal government on scientific and technical matters. Dr. Ralph J. Cicerone is president of the National Academy of Sciences.

The **National Academy of Engineering** was established in 1964, under the charter of the National Academy of Sciences, as a parallel organization of outstanding engineers. It is autonomous in its administration and in the selection of its members, sharing with the National Academy of Sciences the responsibility for advising the federal government. The National Academy of Engineering also sponsors engineering programs aimed at meeting national needs, encourages education and research, and recognizes the superior achievements of engineers. Dr. C. D. Mote, Jr., is president of the National Academy of Engineering.

The **Institute of Medicine** was established in 1970 by the National Academy of Sciences to secure the services of eminent members of appropriate professions in the examination of policy matters pertaining to the health of the public. The Institute acts under the responsibility given to the National Academy of Sciences by its congressional charter to be an adviser to the federal government and, upon its own initiative, to identify issues of medical care, research, and education. Dr. Victor J. Dzau is president of the Institute of Medicine.

The **National Research Council** was organized by the National Academy of Sciences in 1916 to associate the broad community of science and technology with the Academy's purposes of furthering knowledge and advising the federal government. Functioning in accordance with general policies determined by the Academy, the Council has become the principal operating agency of both the National Academy of Sciences and the National Academy of Engineering in providing services to the government, the public, and the scientific and engineering communities. The Council is administered jointly by both Academies and the Institute of Medicine. Dr. Ralph J. Cicerone and Dr. C. D. Mote, Jr., are chair and vice chair, respectively, of the National Research Council.

www.national-academies.org

PLANNING COMMITTEE FOR THE WORKSHOP WHEN SCIENCE AND CITIZENS CONNECT: PUBLIC ENGAGEMENT ON GENETICALLY MODIFIED ORGANISMS[1]

DIETRAM A. SCHEUFELE (*Chair*)*, University of Wisconsin–Madison
JOSEPH ARVAI*, University of Calgary
SARAH DAVIDSON EVANEGA, Cornell University
FRED GOULD* (NAS), North Carolina State University
BROOKE SMITH*, COMPASS

*Members of the Roundtable on Public Interfaces of the Life Sciences.

[1]The workshop planning committee members were not involved in the writing of the workshop summary.

ROUNDTABLE ON PUBLIC INTERFACES OF THE LIFE SCIENCES

Members

CYNTHIA BEALL (NAS; *Cochair*), Case Western Reserve University
DIETRAM A. SCHEUFELE (*Cochair*), University of Wisconsin–Madison
JOSEPH ARVAI, University of Calgary
RICK BORCHELT, US Department of Energy
DAVID E. DUNCAN, Freelance Journalist
DAVID GOLDSTON, Natural Resources Defense Council
FRED GOULD (NAS), North Carolina State University
DAVID INOUYE, University of Maryland
MOLLY JAHN, University of Wisconsin–Madison
BRUCE LEWENSTEIN, Cornell University
MICHAEL LOHUIS, Monsanto
TIFFANY LOHWATER, American Association for the Advancement of Science
GEORGE MATSUMOTO, Monterey Bay Aquarium Research Institute
MATTHEW NISBET, Northeastern University
JOHN OHAB, US Naval Research Laboratory
IVAN ORANSKY, MedPage Today
STEPHEN PALACIOS, Added Value Cheskin
WILLIAM PROVINE, DuPont
VINCENT RACANIELLO, Columbia University
KENNETH S. RAMOS, University of Louisville
BROOKE SMITH, COMPASS
MARY WOOLLEY (IOM), Research!America

NRC Staff

KEEGAN SAWYER, Project Director, Board on Life Sciences
MICHAEL FEDER, Senior Program Officer, Board on Science Education
ZANE MARTIN, Christine Mirzayan Science and Technology Policy Graduate Fellow
LAUREN SONI, Senior Program Assistant, Board on Life Sciences (until January 2015)
HOLLY RHODES, Board on Science Education, Rapporteur
NOMAN GROSSBLATT, Senior Editor

BOARD ON LIFE SCIENCES

Members

JAMES P. COLLINS (*Chair*), Arizona State University
ENRIQUETA C. BOND, Burroughs Wellcome Fund (retired)
ROGER D. CONE, Vanderbilt University Medical Center
JOSEPH R. ECKER, Salk Institute for Biological Studies
SEAN EDDY, HHMI Janelia Farm Research Campus
SARAH C. R. ELGIN, Washington University
DAVID R. FRANZ, Former Cdr USAMRIID; Consultant
STEPHEN FRIEND, Sage Bionetworks
ELIZABETH HEITMAN, Vanderbilt University Medical Center
JOHN G. HILDEBRAND (NAS), University of Arizona
RICHARD A. JOHNSON, Arnold & Porter, LLC
JUDITH KIMBLE, University of Wisconsin–Madison
MARY E. MAXON, Science Philanthropy Alliance
KAREN E. NELSON, J. Craig Venter Institute
ROBERT M. NEREM, Georgia Institute of Technology
MARY E. POWER, University of California, Berkeley
MARGARET RILEY, University of Massachusetts
LANA SKIRBOLL, Sanofi
JANIS C. WEEKS, University of Oregon
MARY WOOLLEY (IOM), Research!America

Staff

FRANCES E. SHARPLES, Director
JO L. HUSBANDS, Scholar/Senior Project Director
JAY B. LABOV, Senior Scientist/Program Director for Biology Education
KATHERINE W. BOWMAN, Senior Program Officer
MARILEE K. SHELTON-DAVENPORT, Senior Program Officer
KEEGAN SAWYER, Program Officer
AUDREY THEVENON, Associate Program Officer
BETHELHEM MEKASHA, Financial Associate
ANGELA KOLESNIKOVA, Administrative Assistant
JENNA OGILVIE, Senior Project Assistant
LAUREN SONI, Senior Project Assistant (until January 2015)

BOARD ON SCIENCE EDUCATION

Members

ADAM GAMORAN (*Chair*), WT Grant Foundation
GEORGE BOGGS, Palomar College (emeritus)
MELANIE COOPER, Department of Chemistry, Michigan State University
RODOLFO DIRZO, Department of Biology, Stanford University
JACQUELYNNE ECCLES, Department of Psychology, University of Michigan
JOSEPH FRANCISCO, Department of Chemistry, Purdue University
MARGARET A. HONEY, New York Hall of Science
SUSAN KIEFFER, Department of Geology, University of Illinois, Urbana
MATTHEW KREHBIEL, Kansas State Department of Education
MICHAEL LACH, Urban Education Institute, University of Chicago
LYNN S. LIBEN, Department of Psychology, The Pennsylvania State University
BRIAN REISER, School of Education and Social Policy, Northwestern University
MARSHALL (MIKE) SMITH, Carnegie Foundation for the Advancement of Teaching
ROBERTA TANNER, Retired Physics Teacher, Thompson School District, Loveland, Colorado
SUZANNE WILSON, Neag School of Education, University of Connecticut
YU XIE, Department of Sociology, University of Michigan

Staff

HEIDI SCHWEINGRUBER, Director, Board on Science Education
MICHAEL FEDER, Senior Program Officer
MARGARET HILTON, Senior Program Officer
MATT LAMMERS, Program Coordinator

Acknowledgment of Reviewers

This workshop summary has been reviewed in draft form by persons chosen for their diverse perspectives and technical expertise in accordance with procedures approved by the National Research Council's Report Review Committee. The purposes of this review are to provide candid and critical comments that will assist the institution in making the published summary as sound as possible and to ensure that the summary meets institutional standards of objectivity, evidence, and responsiveness to the charge. The review comments and draft manuscript remain confidential to protect the integrity of the deliberative process. We thank the following for their participation in the review of this summary:

Richard Amasino, University of Wisconsin-Madison
Kevin M. Folta, University of Florida
Geoffrey Hunt, American Society for Biochemistry and Molecular Biology
Margaret Mellon, independent science policy consultant
Louie Rivers III, North Carolina State University
Allison A. Snow, Ohio State University

Although the reviewers listed above have provided many constructive comments and suggestions, they were not asked to endorse, nor did they see, the final draft of the workshop summary before its release. The review of the report was overseen by Diane Griffin of Johns Hopkins University. Appointed by the National Academy of Sciences, she was responsible for making certain that an independent examination of the report was carried out in accordance with institutional procedures and that all review comments were carefully considered. Responsibility for the final content of the summary rests entirely with the authors and the National Research Council.

Contents

Boxes

Figures

Appendices

1

Introduction and Overview

Rapidly advancing scientific knowledge and its applications from a wide variety of disciplines–genetic engineering, climate science, synthetic biology, stem cell research, and others –are often the subject of multifaceted societal debates. Some of the debates revolve around scientific questions that scientists, as respected authorities, are called on to answer: How fast is the global climate warming? Does the consumption of genetically engineered corn cause allergic reactions? And some of the debates revolve around complex ethical or policy questions to which scientists, as stakeholders and not necessarily authorities, may contribute: Should research involving human embryos be allowed? Should people or institutions be allowed to hold patents on genes?

Given the multifaceted nature of many societal debates about science, how can scientists engage with members of the public to empower decision-making and participation in public policy? That question led the National Research Council's Roundtable on Public Interfaces of the Life Sciences (PILS; See Appendix D) to hold a 2-day workshop on January 15–16, 2015, in Washington, DC, to discuss key components of scientific engagement with the public The workshop presentations and discussions dealt with perspectives on scientific engagement in a world where science is interpreted through a variety of lenses, including cultural values and political dispositions, and with strategies based on evidence in social science to improve public conversation about controversial topics in science.

The workshop focused on public perceptions and debates about genetically engineered plants and animals, commonly known as genetically modified organisms (GMOs), because the development and application of GMOs are heavily debated among some stakeholders, including scientists. For some applications of GMOs, the societal debate is so contentious that it can be difficult for members of the public, including policy-makers, to make decisions. Thus, although the workshop focused on issues related to public interfaces with the life science that apply to many science policy debates, the discussions are particularly relevant for anyone involved with the GMO debate. The Statement of Task for the workshop planning committee is in Box 1-1

Key topics discussed during the workshop included the following:

- The cognitive processes involved in how people evaluate science information and make decisions.
- The information environment that influences public perceptions of science.
- The cultural and political contexts that surround science in general and GMOs in particular.

In his opening comments, Dietram Scheufele, of the University of Wisconsin–Madison and chair of the PILS Roundtable, remarked that application of GMOs, particularly in food and agriculture, have been on the public agenda for a long time; it is not a new issue. He asked, Why are we still talking about GMOs? He offered two primary answers to this question:

- GMO research is a highly politicized field of science, which is true of "most emerging technologies".
- GMO research is post-normal – science for which political systems are uncertain and the decision stakes tend to be high.

BOX 1-1
Statement of Task

An ad hoc committee will plan and convene a public workshop to explore the public interfaces between scientists and citizens (e.g., consumers, farmers, and corporate or government policy-makers) in the context of genetically engineered (GE) organisms. The workshop discussions will explore the empirical findings from social science disciplines on market dynamics, public opinion, attitudes, and decision-making in the US and abroad. Ethical, legal, and other societal value systems of scientists and decision-making audiences that underlie public debates about genetic engineering, and what is known about successful models of engagement given those values will also be discussed. Finally, the workshop will delve into the science information needs of decision-makers, and potential collaborative mechanisms that facilitate access to and evaluation of scientific evidence about GE organisms for decision-making purposes. Some of the questions addressed at the workshop include:

- What values or value systems influence the attitudes of scientists and publics towards genetically engineered organisms?
- How can scientists and science policy-makers enter into dialogue with the public on issues related to genetically engineered organisms in ways that build trust?
- What is the appropriate (and realistic) role of science in informing decisions related to genetically engineered organisms?
- What types and sources of information about genetically engineered organisms are useful and credible to citizens, given their diverse value systems?
- How can scientific information about genetically engineered organisms be best presented for use by policy decision-makers?
- How can non-scientists and consumers access and evaluate scientific studies about genetically engineered organisms in real time, to better inform their decisions?

He said that the characteristics of science and technology that lend themselves to becoming politicized and post-normal are high complexity, fast bench-to-bedside transitions, and ethical, legal, and social issues that are as important as the scientific capabilities.

Scheufele explained that the PILS Roundtable took on the issue of public engagement on GMOs because of an increasing awareness among natural scientists that many emerging technologies in the life sciences, like GMOs, affect society directly. "They may create concerns. They certainly have created lots of policy debates and have influenced market dynamics," he said. Scheufele defined public interfaces of GMOs as any connection of the science of GMOs with societal applications and political effects. He emphasized that when it comes to figuring out how to build better science–society conversations on GMOs, "we spent a lot of time winging it." However, empirical findings of social-science research on public perceptions of science could be used to inform science–public interfaces. Hence, Scheufele outlined the goals of the workshop as investigating: findings from the behavioral and social sciences about how these interfaces work, learning how the public reacts to the different aspects of GMO technology, and discussing how to build empirically based science–public interfaces.

WORKSHOP OVERVIEW

On Day 1, Scheufele opened Session 1 by welcoming participants, outlining the goals of the workshop, and describing a series of myths about public perceptions of science that have influenced how people communicate about GMOs. William Hallman, of Rutgers University, then discussed his research on how consumers make decisions, particularly about GMOs. Dan Kahan, of Yale University, described conditions that can erode what he called "the science communication environment" and factors that might be at play in discussions about GMOs. Finally, Roger Pielke Jr., of the University of Colorado, talked about the role of policy and politics in science and about how cultural and political contexts affect the communication of science.

In Session 2, three speakers discussed knowns, unknowns, and challenges related to public perceptions of science. Dominique Brossard, of the University of Wisconsin–Madison, discussed public perceptions of GMO technologies. Stephen Palacios, of Added Value Cheskin, described market research that food industries use to evaluate public perceptions and why it matters. Food and science journalist Tamar Haspel, and independent science and health journalist, described the challenges of personal biases in science journalism and approaches to overcome them. Jason Delborne, of North Carolina State University, described approaches to public engagement in ways that include publics with diverse perceptions, including opposing perceptions, of GMOs. Session 2 concluded with a panel discussion on the role of science and scientists in public initiatives to label genetically modified foods. Panelists were Robert Goldberg, of the University of California, Los Angeles, Eric Sachs of Monsanto, Allison Snow, of Ohio State University, William Hallman, and Tamar Haspel.

On Day 2, Brooke Smith, of COMPASS, opened with information and insights gleaned from Day 1 presentations and discussions. Workshop participants then separated into three breakout groups to discuss how lessons from the workshop apply to different societal conversations about GMOs, specifically transgenic corn and the monarch butterfly, the American chestnut, and genetically modified mosquitoes. Summaries of the breakout sessions were shared and discussed in plenary session. The workshop concluded with a four-member reaction panel and a facilitated audience discussion about conceptual and practical take-homes from the workshop. The panelists were Rick Borchelt, of the Department of Energy, Helen Dillard, of the University of California, Davis, Molly Jahn, of the University of Wisconsin–Madison, and Dan Kahan.

The workshop was attended by 90 persons, and another 126 joined via webcast. On-line participants were encouraged to ask questions and contribute to discussions via Twitter at #NASInterface. Workshop presentations and archived videos are available through the PILS Web site.[2] The workshop agenda, a list of participants who attended in person, and the biographies of speakers and the workshop planning committee can be found in Appendixes A, B, and C.

ABOUT THIS SUMMARY

This report summarizes the presentations and discussions that took place during the workshop. It is organized by major themes. Written by rapporteurs, this publication is a factual summary of the presentations and discussions at the workshop. The organizing committee took no part in the writing of the summary. The organizing committee extended invitations to a broad spectrum of individuals. This summary represents the views expressed by the individual workshop participants and so is not necessarily representative of all viewpoints. Nor do the views necessarily represent the organizing committee or the National Academy of Sciences. In accordance with the policies of National Research Council, this document does not establish any conclusions or recommendations of the National Research Council; instead, it focuses on issues and ideas presented by the speakers and workshop participants.

[2] http://nas-sites.org/publicinterfaces/

2

How People Think (about Genetically Modified Organisms)

"I want to assure you that the public is not irrational."—William Hallman

We don't think the way that we think we do. How scientist communicate with members of the public is often misguided by many commonly held but erroneous assumptions about how people form opinions and make decisions. Workshop participants discussed evidence from a small, rapidly growing social-science discipline, the science of science communication, which debunks many of the commonly held beliefs. The presentations highlighted results of research on science communication and public perceptions of GMOs and other scientific subjects.

DEBUNKING MYTHS ABOUT PUBLIC PERCEPTIONS OF SCIENCE

Scientists often blame science illiteracy, unscientific thinking, and distrust for societal debate about science and its applications, such as that on GMOs. Speakers Dietram Scheufele of the University of Wisconsin- Madison (UW Madison), Tamar Haspel a food and health journalist with the Washington Post, Dan Kahan of Yale University, Dominque Brossard of UW Madison, addressed these perceptions, myths, and the social science evidence that refutes them.

Myth 1: Knowledge Deficits Are Responsible for a Lack of Public Support of Science

The so-called knowledge-deficit model or familiarity hypothesis has two primary assumptions, said Scheufele. First, "if people were more informed, they would ultimately be more supportive of science"; second, "we need to go out and just get more knowledge in the system and then things will be better for science." The knowledge-deficit model fuels many scientists' desire and efforts to increase the public's science literacy. Scheufele emphasized, however, that social-science research provides evidence that more scientific information or knowledge about an issue does not lead to greater public support of particular scientific findings, and may even produce the opposite effect. Kahan provided an example of that effect from his research on climate change and public literacy.[3] Using National Science Foundation data on public science literacy and technical reasoning capacity, Kahan and colleagues found that when a person's political outlook is more liberal, concern about climate change increases as science literacy increases; however, if one's political outlook is more conservative, concern about climate change decreases as science literacy increases. In other words, increased knowledge was not a driver of public perception of risks associated with climate change. Why doesn't providing people with more information increase their support for scientific evidence on changes in the Earth's climate, the

[3]Kahan, D.M., E. Peters, M. Wittline, P. Slovic, L. L. Ouellett, D. Braman, and G. Mandal. 2012. The polarizing impact of science literacy and numeracy on perceived climate change risks. *Nature Climate Change* 2:732–735.

effects of GM crops on human health, or other areas of research? The answer is explained in part by *motivated reasoning* and *confirmation bias*[4]. Motivated reasoning occurs when information that fits one's beliefs is weighted more heavily than information that does not fit them, Scheufele explained. All people, including scientists, tends to engage in motivated reasoning. Motivated reasoning helps to explain why the same piece of scientific information can have different meanings to different people and in turn produce different outcomes, he stated.

Haspel shared a metaphor for motivated reasoning described in *The Righteous Mind:* the elephant and the rider.[5] The elephant is the "sum total of our intuitions, our emotions, our cultural affiliations, our values, all things that are essentially not quite innate but that are operating below the level of our rational mind," Haspel explained. Atop the elephant is a rider, and the rider represents our cognitive processing that is responsible for analyzing facts and reaching logical conclusions. A person's elephant more quickly decides what it thinks about an issue than the rider, she said. For example, "I love the idea that we will be able to feed more people more efficiently; or 'yuck,' this is a terrible idea—we are taking genes from one species and putting them into another." Once the elephant has decided which way to go, it is difficult for the rider to change the course (Figure 2-1).

Our desire to confirm what we believe also emerges in what social scientists call confirmation bias. We seek out people, mass media, and other information sources that agree with what we believe, Haspel explained. And we tend to filter out information that conflicts with what we believe. People can also hear disconfirming information and respond by becoming more entrenched in their positions, she said. Finally, people tend to evaluate the credibility of experts according to the extent to which they agree with them. Confirmation bias "is pretty scary because it basically allows us to believe the only credible people are the ones who share our worldview. It becomes difficult to find a way to change your mind," Haspel said.

All people have differing experiences, characteristics, beliefs, and values that affect how they weigh scientific information. Scheufele, Kahan, and Brossard all emphasized that motivated reasoning and confirmation bias can be seen in societal debates about many issues, such as the use of stem cells, climate change, fracking, and gun control. To illustrate that point, Scheufele described results of a study on the effects of religious and other personal values on public attitudes about embryonic-stem-cell research.[6] Among the highly religious participants in the study, people who had more knowledge about embryonic stem cells were no more supportive of stem-cell research than people who had less knowledge. "It is not about their not knowing, and it is not about their not getting the science," said Scheufele. Rather, what they do know they don't necessarily translate into a more favorable attitude toward stem cell research, he concluded.

What about GMOs? To set the stage, Brossard shared data from public-opinion polls over the last decade, which consistently show that roughly 10% of the American public favors the use of biotechnology in plants that produce food "not at all", and another 10% finds it "very favorable." In other words, there are "very vocal minorities, pro and con" that hold strong views, she explained. Roughly one-third of the American populace "does not care"— does not have sufficient time, energy, or interest to invest intellectual energy in the GMO debate, Brossard stated. Next, she noted that results of CBS/New York Times/60 Minutes/Vanity Fair[7]

[4] Kunda, Z. 1990. The case for motivated reasoning. *Psychological Bulletin* 108(3):480-498.

[5] Haidt, J. 2013. *The righteous mind: Why good people are divided by politics and religion.* New York, NY: Vintage Books.

[6] Ho, S. S., D. Brossard, and D. A. Scheufele. 2008. Effects of value predispositions, mass media use, and knowledge on public attitudes toward embryonic stem cell research. *International Journal of Public Opinion Research*, 20(2):171-192.

[7] CBS News, The New York Times, 60 Minutes, and Vanity Fair. CBS News/New York Times/60 Minutes/Vanity Fair National Poll, June #1, 2012. ICPSR34642-v1. Ann Arbor, MI: Inter-university Consortium for Political and Social Research [distributor], 2013-06-05. http://doi.org/10.3886/ICPSR34642.v1.

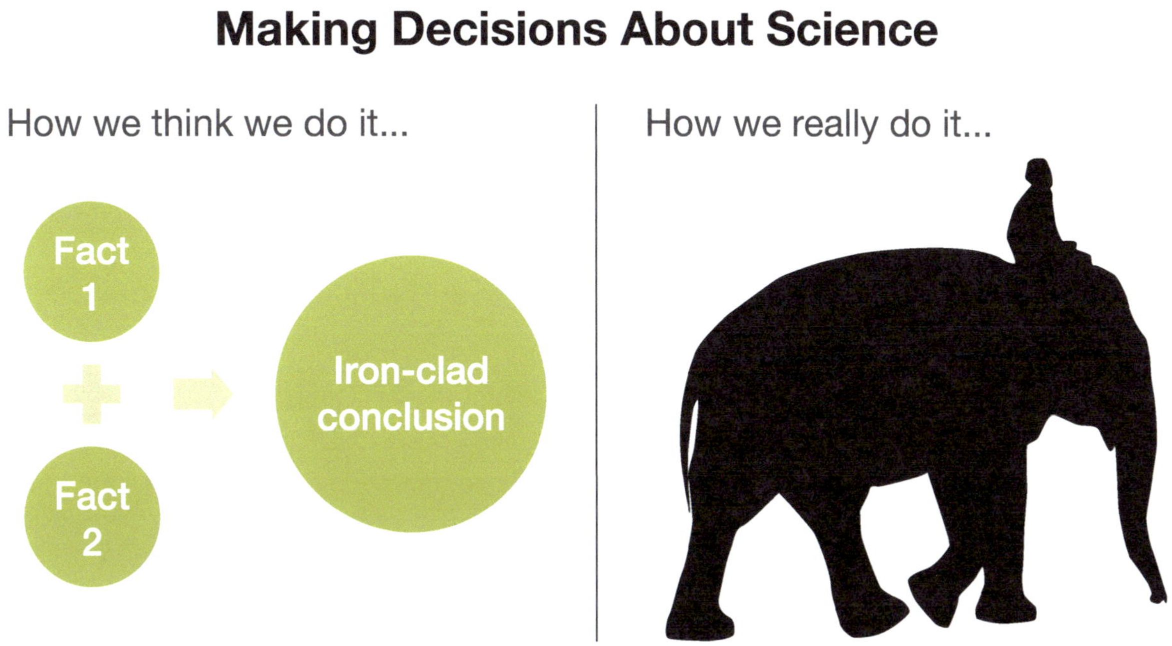

Figure 2-1 Making decisions about science. Tamar Haspel used the metaphor of the elephant and the rider to describe how people make decisions. Although many believe that they make decisions analytically and rationally (the rider), in fact people usually make decisions based on their underlying cultural values (the elephant). People often find it far easier to justify why their elephant chose the correct path(decision) than to change its course. SOURCE: Haspel, workshop presentation, slide 4.

and CBS/New York Times [8] national public-opinion polls in June 2012 and January 2013, respectively, show that of roughly two-thirds of the what concerns people most about GMOs in the United States are related to health (for example, risks of cancer or food allergies) and on whether genetically modified food is safe to eat and that "knowledge level accounts for a small amount of the variance in public attitude." In short, risk perception is not determined by people's knowledge about potential hazards, Brossard explained. She added that people tend to assess risks posed by technologies, such as GMOs, on the basis of a combination of knowledge about the hazards and benefits *and* their potential to disrupt our lives in a manner that provokes personal outrage.

David Goldston, of the Natural Resources Defense Council, cautioned participants not to "overcorrect" for the deficit model. He emphasized that facts are important in political debates about scientific issues. "The idea that facts are irrelevant, I think, is a gross overstatement and a dangerous one. What makes it tricky," Goldston said, "is that we do not know when and how and which facts are going to matter."

Myth 2: Useful Public Debate Requires Citizens That Think Like Scientists

What does it mean to think like a scientist? We tend to believe that "scientists think logically, analytically, and dispassionately and that members of the public are more intuitive and emotional," Kahan said. He added that scientists tend to believe that the public "overestimates more dramatic risks like terrorism" as opposed to more remote risks like species extinction due to climate change. "As people become better at the kind of reasoning that is characteristic of science, they don't converge on what scientists say, they just become better

[8] CBS News and The New York Times. CBS News/New York Times National Poll, January #1, 2013. ICPSR34991-v1. Ann Arbor, MI: Inter-university Consortium for Political and Social Research [distributor], 2014-04-01. http://doi.org/10.3886/ICPSR34991.v1

indicators of what people like them think about this issue," Kahan explained.

Kahan discussed the extent to which orchestrated misinformation might explain why there is societal debate about science. That assumption posits that economically motivated interest groups are supplying misinformation to a credulous public. In fact, what appears to be the case, according to Kahan, is that people are misinforming themselves and are not using new evidence to update their beliefs and understandings. "You have a culturally motivated public that really is eager to find information that is consistent with what their group believes. They will even use their science comprehension and critical reasoning skills to do it." The result of that motivated reasoning is that they create a demand for misinformation, Kahan said.

The irony of the myth that training citizens to think more like scientists would improve societal debate about science is that "scientists don't think like scientists," particularly with regard to ethical, legal, or social implications of their work, Scheufele stated. To illustrate that point, Scheufele described research results from regularly conducted surveys and interviews with leading scientists about the societal and public policy interfaces of their research, such as risks, benefits, and the need for regulation.[9] An initial review of the findings revealed that scientists believe that the greater the risks of a new technology, such as nanotechnology, the greater the need for regulation. However, further analysis demonstrated that the scientists' personal ideologies predicted their stances on regulation, even after statistically controlling for their field of expertise and their seniority in their field. "The idea that scientists can ask members of the public to think scientifically about the political or social implications of technology is naive because we, as scientists, do not do it either," Scheufele concluded.

So how do we all think? People are *cognitive misers,* Scheufele said. We use *mental shortcuts* to process information. Beliefs—ideologies, values, partisanship, and others—serve as mental shortcuts, ways to judge information quickly, he said. Because "it is not possible to use all information available to make [all our] decisions," we rely on *low-information rationality;* that is, it is rational for people not to seek all available information but to rely on beliefs, he said[10]. Brossard emphasized that mental shortcuts are healthy human responses to the multitude of decisions that need to be made.

Myth 3: The Public Does Not Trust Scientists

"Trust matters more than knowledge" with respect to whom people listen to about science, Brossard asserted. That is, people accept the message when they trust the messenger. She added that the dimensions by which a group is assessed as trustworthy vary. For example, businesses must build trust that they care and are paying attention to people, advocacy groups must build confidence in their knowledge and expertise, and government has to establish that it is honest and open, she explained. "Trust does not mean the same thing for everybody," she summarized.

Some have argued that the root of problems in societal debates about science is a declining trust in scientists, Scheufele noted. However, he pointed out that data from the General Social Survey[11] that demonstrate the percentage of Americans that express "a great deal of confidence" in the scientific community has remained roughly unchanged since the early 1980s (Figure 2-2). Scheufele's group's research on societal perceptions of nanotechnology indicates that university and industry scientists are considered among the most trusted sources of information on the technology.[12]

[9] Corley, E. A., D. A. Scheufele, and Q. Hu. 2009. Of risks and regulations: How leading US nanoscientists form policy stances about nanotechnology. *Journal of Nanoparticle Research*, 11(7):1573-1585. doi: 10.1007/s11051-009-9671-5.

[10] Scheufele, D. A. 2006. Messages and heuristics: How audiences form attitudes about emerging technologies. Pp. 20-25 in J. Turney (ed.), *Engaging science: Thoughts, deeds, analysis and action*. London: The Wellcome Trust.

[11] The General Social Survey (GSS) has gathered data on contemporary American society—trends, attitudes, behaviors, and attributes—since 1972. The GSS project is run by NORC of the University of Chicago. http://www.norc.org/Research/Projects/Pages/general-social-survey.aspx.

[12] Scheufele, D. A., E. A. Corley, T.-J. Shih, K. E. Dalrymple, and S. S. Ho, S. S. 2009. Religious beliefs and public attitudes to nanotechnology in Europe and the US. *Nature Nanotechnology*, 4(2): 91-94. doi: 10.1038/NNANO.2008.361.

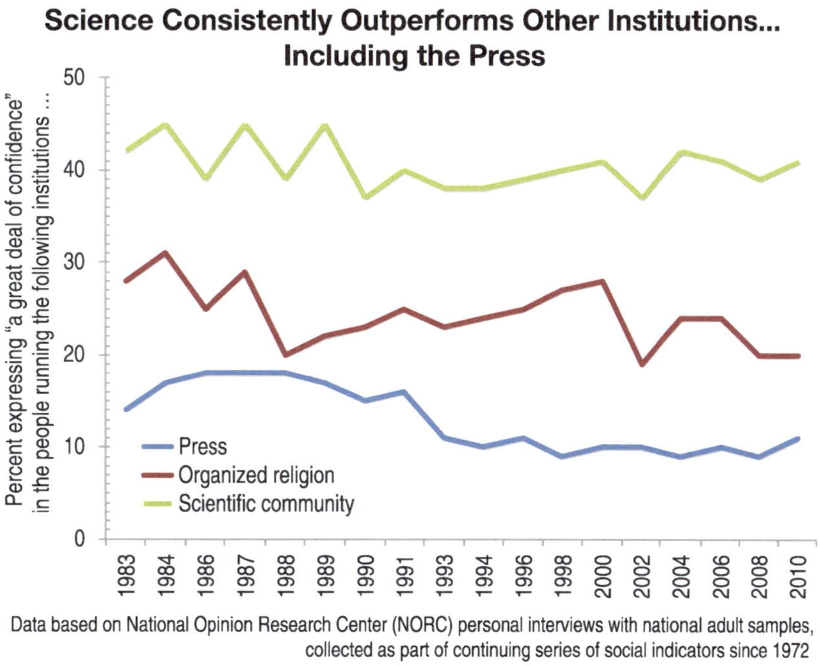

Figure 2-2 American's Trust in the Scientific Community, 1983-2010. Drawing from GSS survey data, the graphs compare trust in the press, trust in organized religion, and trust in scientists. American trust in the press and organized religion is lower and in decline, and trust in science is relatively higher and stable. SOURCE: Scheufele, workshop presentation, slide 15.

Kahan has also studied whether there is growing distrust of science or scientists. He explored National Science Board (NSB) Science and Engineering Indicator results on American measures of attitudes toward science, of whether people agree that the federal government should fund science, and of whether people think that the government funds scientific research too little or too much.[13] Based on NSB data, Americans think that scientists are public-spirited, help to solve challenging problems, and are trying to do good on behalf of society. Kahan and his colleagues also examined the degree to which people trust scientists by presenting people with profiles of people who are experts in particular scientific issues.[14] They measured participants' cultural outlooks. Current theory predicts that how people form their ideas about risk and strengthen their commitment to groups will reflect their cultural outlooks. Knowing that people differ on those dimensions, they presented half the study participants with profiles of scientists who took positions that indicated high risk and half with profiles that indicated low risk—that is, climate change is certain, is human-caused, and has dire consequences vs it's too early to say, and the models are inconclusive. They found that participants tended to consider someone an expert that they would listen to if the person's position was consistent with their own cultural outlook; otherwise, they dismissed the views of the person. However, none of the study participants believed that their views were inconsistent with the scientific consensus. Kahan concluded that people trust scientists, "but they are motivated to see that what science says is consistent with what their group says."

THE ROLES OF THINKING AND FEELING IN DECISION-MAKING (ABOUT GENETICALLY MODIFIED ORGANISMS)

William Hallman of Rutgers University described how people, in making different types of decisions, are guided by their worldviews, social

[13]http://www.nsf.gov/statistics/seind14/

[14]Kahan, D. M., H. Jenkins-Smith. and D. Braman. 2011. Cultural Cognition of Scientific Consensus. *Journal of Risk Research* 14:147-174.

comparisons, and concern with self-presentation. Hallman explained that people often overestimate how much they know about a subject, decreasing the likelihood that they will seek more information, but still form opinions that only become more ingrained as they act on them. He suggested that what some scientists want to know is whether members of the public can reach the *right* decisions about GMOs. “My answer is yes, but it depends on what your definition of right is.”

Decisions about What?

Hallman said that the first step in understanding how consumers make decisions is to ask, Decisions about what? What kind of decisions? He explained that people make three basic types of decisions: decisions about truthfulness of facts, evaluative decisions, and behavioral decisions. To examine how people make decisions about truthfulness, Hallman and his colleagues conducted surveys to examine consumers’ beliefs about GMOs.[15] They asked 1,148 adults selected from a representative panel of American consumers to rate the truthfulness of claims about GMOs commonly found on the Internet on a scale ranging from definitely true to definitely false or “I don’t know”. Nearly 70% of the survey participants did not know whether the claims were true or false, Hallman said. The survey results demonstrate that people have a great deal of uncertainty and are unable to detect whether claims about GMOs are factual, he stated.

Evaluative decisions require people to weigh the risks and benefits associated with GMOs, Hallman explained. Even though many Americans do not have a good understanding of what *GMO* means or what the technology entails, they still make evaluative decisions about it, he said. For example, many people who respond that they are not in favor of using genetic modification to create new kinds of plants respond in the same survey that they would be in favor of creating trees that can clean contaminated water or of creating more nutritious grain to feed people in developing countries. Decisions that require people to determine how consistent a technology, another person, or a corporation is with their values, worldviews, or ideologies involve another type of evaluative decision, Hallman explained.

Behavioral decisions are about whether to act. Voting on a referendum to label genetically modified foods, buying a product because it is labeled “GMO-free”, and protesting a local supermarket that carries genetically modified corn are behavioral decisions, Hallman noted.

People use several overarching types of mental shortcuts to make all three types of decisions, Hallman said. One is their general worldview, such as their views on “naturalness” or beliefs about the fallibility of humans. Another involves making social comparisons on the basis of group norms; this involves considering what decisions about an issue others have made to determine what to think or do. Finally, people are affected by their desire to be seen in particular ways to fit into a social or cultural group; this shortcut focuses more on self-presentation that is consistent with the group than on the “truth” of an issue. Motivation and the ability to digest and consider factual information influence how much one relies on mental shortcuts, according to Hallman.

Reality Filters

People have different perceptions about science and therefore often reach different decisions related to science. Hallman explained that people generally overestimate how representative their own knowledge and beliefs are of those of the whole population. When they see that others have reached different conclusions, the natural human response is to either question others’ competence or motivation. Hallman indicated that although that is a human tendency, the idea is especially reinforced in science. The scientific method itself is based on the idea that if assumptions, methods, and data are explained, the same conclusions should be reached. When they are not, assumptions, methods, and data are re-examined. However, that is not how people think in their day-to-day lives, Hallman suggested: “One of the things I often hear from my science colleagues is that the public is certainly just irrational. But I want to assure you that the public is not irrational. They actually do have a basis for their decisions. They are

---

[15]Hallman, W. K., C.L. Cuite, and Z. K. Morin. 2013. Public perceptions of labeling genetically modified foods: Working Paper 2013-01. Rutgers School of Environmental and Biological Sciences. http://humeco.rutgers.edu/documents_PDF/news/GMlabelingperceptions.pdf.

BOX 2-1
Four Filters of Reality

Literacy: the ability to understand the meaning of words and stories.
Graphicacy: the ability to learn graphic (visual) information.
Numeracy: the ability to understand numbers.
Ecolacy: the ability to understand complex relationships.

Literacy + Numeracy + Graphicacy ≠ Ecolacy

Source: Hallman, workshop presentation, slides 9 and 15.

just different from the ones that natural scientists have."

All people, including scientists, Hallman said, use four filters of reality: literacy, graphicacy, numeracy, and ecolacy (Box 2-1).[16] He explained that most people learn and communicate through literacy—stories, anecdotes, metaphors, and so on. However, he noted that the ability to take meaning from words and stories is culture-specific. Cultures often have their own "kinds of myths and kinds of references points" that may not be understandable to people in other cultures. Graphicacy—as related to sketches, photographs, diagrams, maps, and other forms of visual imagery—is also often culture-constrained, Hallman said. For example, the skull and crossbones represents danger to some, but child-poisoning studies demonstrate that the image doesn't intrinsically indicate poison.

Scientists often use a numeracy filter in their research. Laypeople may struggle more with mathematical concepts, particularly with very large and very small numbers, fractions, proportions, percentages, and probabilities, Hallman explained. When scientists communicate through numbers while journalists and other nonscientists communicate primarily through words and images, a communication barrier can arise. Numeracy barriers can exist even between scientific disciplines, he added.

[16]Hardin, G. (1985). "The Expert as Enemy and Three Filters of Reality." Pp. 7-25 in G. Hardin, *Filters Against Folly: How to Survive Despite Economists, Ecologists, and the Merely Eloquent*. New York: Viking Penguin.

The ability to see "the big picture" and the capacity to envision intended and unintended consequences of a decision or action are encompassed by ecolacy. Hallman implied that the ability to have a conversation about GMOs requires ecolacy within the scientific community and within diverse publics. However, he cautioned, being skilled in literacy, numeracy, and graphicacy does not equate to having ecolacy: "Just knowing a string of facts does not give you the ability to put them together to see the whole." And that is why "educating people about the scientific details does not necessarily lead to greater comprehension of the big picture or the ability to make informed decisions."

DECISION-MAKING ABOUT GENETICALLY MODIFIED ORGANISMS

Hallman explained that all decisions have *cognitive* (thoughts) and *affective* (feelings) components. How people combine cognitive and affective components depends in part on the types of decisions that they are making. According to Hallman, many psychologists and economists believe that affect is a by-product of cognition; that is, people's evaluation of information leads to an emotional response. In fact, the opposite is often true, and this explains in part why first impressions matter so much. An initial emotional response affects later thinking. The pattern can also be cyclical: "I like it because it is good, and it is good because I like it." People often have a poor understanding of what influences their perceptions and behaviors and of why they feel, choose, and act in particular ways. Furthermore, people often cannot predict what will make them happy in the future. In sum, it is erroneous to believe that thinking always comes before feeling, according to Hallman.

Findings from Hallman's 2013 survey demonstrate those complex relationships in Americans' ideas about and perceptions of GMOs. One set of questions asked respondents to rate their knowledge of how food is grown and produced in the United States and then presented a set of factual questions to measure their actual understanding. Although only 22% of participants rated their understanding as fair or poor, responses on the knowledge test indicate that people tend to overestimate how much they know. A consequence of the overestimation is that if people believe that they already know about a topic, they are unlikely to seek information or ask questions, Hallman explained.

The survey also revealed that 25% were not aware that genetically modified foods existed before they took the survey, 55% rated their own knowledge of genetically modified foods as little or none, and 66% had never had a conversation about GMOs with anyone. In addition, only 43% knew that there are genetically modified foods in supermarkets today, and only 25% believed that they were eating genetically modified foods. In other words, the survey results demonstrate that people "have not heard very much, they do not know very much, they have never talked about it, they are unaware that they are eating it, and yet they have an opinion," Hallman summarized. In fact, about 50% of those surveyed admitted that their opinion of genetically modified foods was based on their "gut feeling" about it. Only 15% stated that their opinion was based on a specific issue, and 34% said that their opinion was based on both gut feeling and a specific issue. Hallman emphasized that "being uninformed doesn't stand in the way of having an opinion." Most of the uninformed opinions are not strongly held, so they are subject to change. Nevertheless, they matter because people make decisions and act on the basis of them, he added.

Science issues involving food carry particular significance, and they are different from other science issues about which people make decisions, according to Hallman. Food is taken internally, and it also has symbolic value for nurturing and health and plays an important role in our relationships with others. Food is particularly susceptible to the "yuck" factor, the mental shortcut of disgust, he stated. Environmental technologies and even medicines, which are also taken internally, do not face the same challenges. This framework leads to the type of mental shortcut that can drive decision-making, Hallman concluded.

3

The Science-Information Climate

"There is no unframed information."—Dominique Brossard

The scientific community has one of the many voices involved in public discussions about the societal applications and implications of GMOs. In addition to research on how a person thinks and makes decisions, social-science research focuses on the *information climate* that influences public opinions and societal discussions about science. Workshop presenters discussed both broad characteristics of the information climate and the *publics,* the groups of people that participate in societal conversations about science and technology.

Dominique Brossard of the University of Wisconsin-Madison (UW Madison) described influences on public perceptions of science as having three layers: individual characteristics, the information climate, and sociopolitical and cultural contexts (Figure 3-1). Often, people focus only on individual characteristics, such as how the political ideology of one citizen may influence how that person perceives biotechnology, Brossard said. Indeed, research on how people think and make decisions has led to the finding that mental shortcuts (Chapter 2) influence a person's perception of science more strongly than knowledge about the science. However, she underscored that the information climate—including messages from the mass media, commercial marketers, the entertainment industry, and schools—often shapes the mental shortcuts that people use.

FRAMING SCIENCE INFORMATION

"We think that we have some form of communication monopoly when it comes to new technologies—that if science speaks, the different publics will listen—and that is not the case," said Dietram Scheufele of UW Madison. The assumption that the public is listening only to scientists about scientific issues is rooted in the deficit model, he stressed. The deficit model posits that people lack information and that when they gain more information from the scientific community, they will make better decisions that are based on their new knowledge. With that one-way science-to-public communication model being largely debunked by social-science research, others have worked to develop bidirectional conversations between scientists and members of the public. The underlying assumption of this approach is that the problem is with how scientists communicate and that ultimately this conversation will lead to a more effective way to tell people what will work better. However, according to Scheufele, many voices are competing to be heard: "Very often, we may not be perceived as the most credible voice unless we manage to position ourselves well in that overall [information] environment."

Scientists lack a monopoly on communicating about new technologies, such as GMOs, in part because various groups can be effective in quickly and simply framing their viewpoints. Frames like visual imagery, headlines, or labels (such as "Frankenfoods") can be quickly and easily understood because they play to perceptions that already exist in people's minds. The concept of *framing*, originally described by Nobel laureates Daniel Kahneman and Amos Tversky, is that all information is reference-dependent and is based on the beliefs that people hold, Scheufele said. People view complex science issues, such as GMOs, in many ways, said Scheufele. He emphasized that framing is an important tool that helps people to make sense of ambiguous information. "Frames help

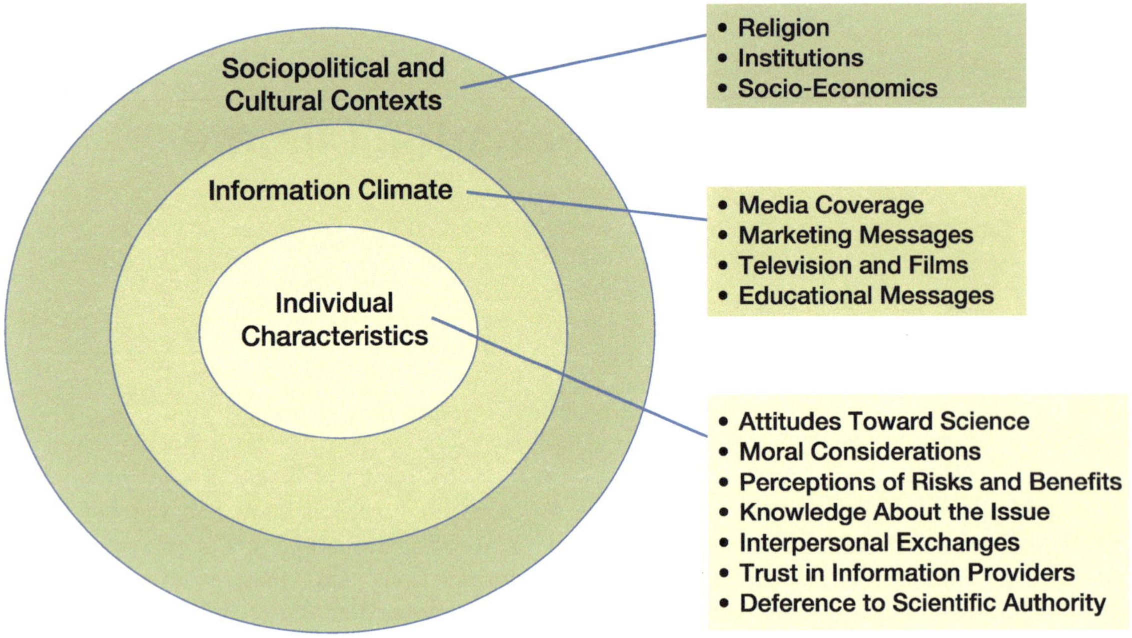

Figure 3-1 Tiers of influence on public perceptions of science. SOURCE: Brossard workshop presentation slides 6, 9, and 10.

us to determine why an issue matters and to process highly complex information," he stated.

"There is no unframed information," Brossard stressed. In addition, the most powerful frames tend to be negative, she said. Still, a given frame can mean different things to different people because of how each person filters communication. How people react to a frame depends on who they are and on their perceptions of the person who shared information with them.

Frames can have unintended consequences, Scheufele explained, as in the case of the 1999 *Washington Post* headline, "Biotech versus the Bambi of Insects: Gene-Altered Corn May Kill Monarchs."[17] Framing the issue that way activated the existing image that most people have of the baby deer Bambi and quickly portrayed biotechnology in a negative way. Such frames are difficult to eliminate, and people may question the motivation of those who try to reframe them, Scheufele said. Because it is so challenging to "unframe" an issue, ideally scientists should think early on about how they will present their science in such ways that laypeople can attach the new information to what they already know, he concluded.

In the context of genetic engineering, many stakeholders contribute to the information climate through a variety of frames, including the mass media, policy-makers, scientists, research institutions, extension staff, farmer groups, industry, activity groups, and consumer groups, Brossard noted. Interpretations are affected by people's culture, political dispositions, interest in science, world views, the source of the message, and a host of other factors.

The words that people use and associate with new technologies both create and shed light on the way in which new technologies are framed. William Hallman of Rutgers University discussed the particular associations that people mention when they think of GMOs and the effects of using particular words to describe them. Some of his research has involved asking people the first things that came to mind when he used various terms, including *GMO*, *biotechnology*, and *genetic engineering*. He found that the term used for the technology was important in determining the types of responses that people gave to the question. He examined whether people named objects or products or whether they provided responses linked to

[17] Weiss, R. 1999. "Gene-Altered Corn May Kill Monarchs," Washington Post, 20 May 1999, Page A3.

emotions. The term *agricultural biotechnology* preferred by the Food and Drug Administration evokes the most neutral responses associated with stereotypical images of scientists. However, *genetic engineering* and *genetic modification* lead to many more negative responses, such as "mutant", "monster", and "nasty".

MEDIA EFFECTS ON THE INFORMATION CLIMATE

The information climate is influenced through several media effects. First, scientific information can be framed in different packages, Brossard explained.

A second mass-media effect is what communication researchers call a *spiral of silence*. It occurs when a vocal minority receives increasing attention in the mass media. That attention is amplified as people talk about what the press is reporting. She said that ultimately people begin to believe that the minority opinion represents a majority opinion— in effect, it silences anyone who does not share the minority opinion.

Cultivation is another mass-media effect that has an impact on perceptions of science, Brossard said. "The more we watch entertainment media, the more we are embedded in a specific portrayal of reality that people end up thinking is the true one," she stated. For example, the extensive amount of violence on television leads people to believe that the world is a meaner place than it actually is, she explained. Scheufele also discussed the powerful effects of cultivation as related to science, stating that most people do not interact with scientists or spend time in laboratories. Therefore, often their only images of scientists come from how they are portrayed in the mass media. People's ideas of nanotechnology come from such movies as *Terminator 3*, and their images of scientists come from such television shows as *The Big Bang Theory* or such movies as *Back to the Future*. Scheufele has found that those effects persist even among science majors at his university—a demonstration of how powerfully the cultivated images shape perceptions.

Each mass-media effect exists within a broader information climate, Broussard stated. The broader climate is shaped by a wide array of stakeholders, including the mass media. In the context of GMOs, policy-makers, press offices in academic institutions, agricultural associations, industry, and consumer groups are among the actors shaping the information climate. Individuals in each also shape the debate as they converse with friends, family members, and colleagues. Mass-media effects within and among those groups are also important to consider when developing approaches to public engagement, according to Brossard.

CONSUMER OPINIONS

Stephen Palacios of Added Value Cheskin discussed the role of consumer opinion about GMOs in forming industry marketing strategies and business decisions. From a marketing perspective, the debate around GMOs has already been framed negatively, Palacios asserted. Evidence of that can be seen simply by conducting an Internet search on the term *GMO*, he said. In a quick Web search, Palacios found that Google results yielded more Web sites against GMOs than in favor of them. Similarly, Palacios found that available movies and books on the topic of GMOs are largely from an anti-GMO perspective. That suggests that the average person who looks for information about GMOs will be presented with negative associations, Palacios said.

Anti-GMO groups have communicated thoughtfully and effectively, Palacios stated. Web sites are often sophisticated and multimodal with both short-form and long-form narrative portions and video. They may have components that appeal to various learning styles and include the ability to engage in "click advocacy" by facilitating direct donations to a specific anti-GMO cause. Some Web sites facilitate registering votes against particular companies through another form of click advocacy, he explained. Other sites, such as Netflix, include the title *GMO OMG*, an award-winning foreign film. Palacios opined that the title and style of the film target the millennial generation, and these forms of communication suggest a level of sophistication and an understanding of the messages' targets.

Palacios indicated that although anti-GMO Web sites appear to have had little effect on the current application of genetic-modification technology in food, some evidence suggests that some food industries are changing their behavior. For example, Chipotle, which had the most successful quick-service restaurant initial public offering in the last 20 years, recently became the first restaurant to label GMOs and develop a plan to go non-GMO in the future, he explained. The Chipotle brand focuses on food integrity, so it has the potential to signal a trend in consumer interests that could influence others, he emphasized.

Food manufacturers and food-industry executives are influenced by consumer opinion regardless of whether the science demonstrates that genetically modified foods are safe, Palacios stated. To illustrate that point, he presented redacted data from a nationally representative consumer study that were shared with the chief marketing officer of a large food and beverage company. The survey focused on determining what consumers want with regard to natural and organic food and beverages and included a set of questions specifically about GMOs. Palacios stated that the summary headlines of the results of the consumer study include such statements as these:

- "Four out of ten consumers today are avoiding or reducing GMOs in their daily diet."
- "GMOs have become potent symbols of the ills of the American food industry."
- "Regardless of organic usage, all consumers express concern about the impact of GMOs on their health."

Palacios noted that the concern over losing relevance and consumers is leading some food industries to have serious discussions about GMOs—whether to consider alternative sources of ingredients, labels, and public-relations activities that might need to occur if consumers become increasingly anti-GMO.

Palacios emphasized that at times both consumers and industry leaders have to make decisions before the science is conclusive. To major industries, consumer perception and opinion are the most important factors, and the fear of losing trust or relevance can drive what and how products are taken to market, a point also expressed by Hallman. Palacios suggested that the science community should focus conversations on targeted applications of GMOs more than working to change wholesale opinions.

Daniel Kahan of Yale University expressed skepticism about whether the results of the industry market survey discussed by Palacios are predictive of consumer behavior. Palacios responded that performance of market-research firms indicates that they are providing value to the industries that they serve. He reiterated that the reality of businesses is that they need to be prepared for changing consumer sentiment, although he did note that information from such surveys is merely taken into consideration in broader strategic discussions. Kahan cautioned that some people want to anticipate public reaction and start a debate where it is not currently happening and that this can have adverse effects on science communication.

ENGAGING PUBLICS IN THE INFORMATION CLIMATE ON GENETICALLY MODIFIED ORGANISMS

As several presenters noted, information that people consume is framed, but the meaning of the frames can be interpreted differently by different groups of people. Delborne of North Carolina State University addressed the notion of who the "the public" is and what this means for the science information climate.

The public is something of a misnomer. Delborne suggested that the plural term *publics* is more appropriate because it "acknowledges that out there in the world there are many different groups of stakeholders, interest groups, people with different accesses to information, different opinions, and so on." However, he argued that *audiences* might be an even more fitting term.

"Audiences are created. The group of publics isn't just out there waiting to be discovered. We construct publics or construct audiences when we attempt to engage them"—a concept that Delborne described in his 2011 publication *Constructing Audiences in Scientific Controversy.*[18]

Delborne discussed the results of a small scale unpublished study that he and his colleagues conducted on public perceptions of genetically engineered mosquitoes. The research team conducted door-to-door interviews with residents of Key West, Florida, in January 2013. First, interviewers explained the genetically engineered mosquito technology. Next, they asked the residents open-ended questions about the perceived hazards and benefits of introducing a genetically engineered

[18]Delborne, J. A. 2011. Constructing Audiences in Scientific Controversy. *Social Epistemology* 25(1):67-95.

mosquito to reduce the population of the mosquito species that are capable of transmitting dengue fever. Results of the study showed 60% of respondents in favor of introducing the genetically engineered mosquito, 23% opposed, and 17% neutral. Those results have different meanings to different people, Delborne explained. For example, Oxitec, the developer of the technology, might react positively to the findings but wonder how it will reach the 23% percent. Alternatively, anti-GMO advocates may be concerned about the findings and wonder, how do we define our message in a better way to make people more concerned? However, Delborne insisted what the results of the study demonstrate is the power and the superficiality of such measures of "support of" and "opposition to" a technology. "The public is constructed in this survey," he said, in terms of who was home when the interviewers knocked on the door, what they were asked, what sort of information they were given, what they expected the results to do, and whether they thought that they were going to be affected by the mosquitoes. Delborne suggested that participatory forms of engagement are better mechanisms to learn about public perceptions of science and how the public navigates the information climate.

Delborne explained that there are three basic types of public engagement—public communication, public consultation, and public participation (Box 3-1).[19] He emphasized public participation means that both members of the public and the scientists might be moved as a result of communicating. The openness of public participation requires both parties to accept some risk, he added.

The consensus conferences developed by the Danish Board of Technology (DBT) are an example of public participation, Delborne said. Experts and lay audiences interact at these conferences with the overt acknowledgment that facts and values are intertwined and inseparable. The purposes of the conferences are to promote learning through deliberation; to develop more thoughtful public opinions; and to generate new ideas and policy alternatives and affect governance decisions. "If you let people deliberate about an idea with good information and you ask them what they think about it, you get a kind of result that is different from what you get if you ask them briefly at their door what they think about GE mosquitoes," Delborne asserted.

BOX 3-1
Three types of public engagement

- Public communication: one-way with information flowing from sponsor to publics in the form of education or outreach.
- Public consultation: one-way with information flowing from publics to sponsor through such mechanisms as opinion polls.
- Public participation: two-way flow of information between sponsor and publics.

SOURCE: Based on Delborne, workshop presentation, slide 11

Delborne shared several lessons learned about useful ways to conduct public engagement through his experiences at two consensus conferences, the 2008 National Citizens' Technology Forum and the World Wide View on Global Warming organized by DBT. First, he found that it is possible to engage in high-quality deliberations with members of various publics. Second, framing the task and questions is important. During the World Wide View on Global Warming, the task was not framed as a debate about whether climate change was real but was bounded in terms of policy options that could be considered at the 2015 Copenhagen Climate Conference. Third, the logistics of constructing publics has an effect on who is in the room. For example, the amount of time that people were asked to use to take part in the National Citizens' Technology Forum meant that the organizers had to offer a substantial stipend. That had the effect of attracting people for whom the stipend was substantial. The composition of the group helps to determine how seriously its deliberations may be taken. Fourth, Delborne noted that "some people are very willing to discount these types of engagement mechanisms because we don't have a perfectly representative sample in the room" even though they will accept the results of an election in which not everyone votes. Delborne' fifth lesson was that empowering participants in public engagement requires skilled facilitators. Facilitators must figure out the degree to which to empower

[19] Rowe, G., and L. J. Frewer. 2005. A Typology of Public Engagement Mechanisms. *Science, Technology and Human Values* 30(2):255.

participants to shift the agenda, decide what to discuss, and decide which experts they want to interact with, Delborne explained. "You risk not only being pulled in your position but being pulled off your agenda and pulled off what you want to talk about." Sixth, for the ideas, decisions, or questions raised in public participation to be useful, "we have to find ways to connect them to real decision-making processes," Delborne said. That brings to the forefront the tension between democracy and expertise, "two values that we hold very high in American society". For many, "turning over some decision-making power to a public that might not align with experts' views is scary and difficult," he concluded.

4

Cultural and Political Contexts

"Sometimes science plays a very small role in the decisions that we make involving scientific topics."—Roger Pielke Jr.

"Let's think in terms of the sociopolitical cultural context in which this debate is taking place before coming up with general conclusions and assumptions about what to do or not do or why and how we should engage the public."—Dominique Brossard

Cultural and sociopolitical contexts are an overarching dimension of what shapes the science-communication environment. During the workshop, presenters and participants discussed the effects of different intersections between science (research) and society (cultural and political contexts) on perceptions of science, and they talked about the array of communication roles that scientists can play. A moderated panel also discussed views on the interfaces between science, culture, and politics, using the specific case of labeling of GMOs in foods.

CULTURAL CONTEXTS

GMOs are a complex, "multidimensional issue that goes beyond food and environmental safety," said Dominque Brossard of UW-Madison. Because societal debates about GMOs have not only technical aspects but ethical, legal, and social dimensions, the subject is considered "controversial". Brossard briefly listed some of the cultural and sociopolitical questions surrounding GMOs (Box 4-1).

Culture plays an important role in determining how an issue is defined, including its risks and benefits, Brossard said. She emphasized that the sociopolitical and cultural contexts differ in different areas of the world, so the concerns about the adoption of genetic modification technology differ. For example, in countries throughout Africa and Asia, more concerns were raised about regulatory mechanisms to ensure that cities were adequately protected, she said. In Europe, concerns about who owns the technology and the effects on local farmers are important. International trade, consumer choice, labeling, and food safety are concerns that vary with location and cultural significance, Brossard explained. In some countries, the societal effects of

BOX 4-1
Examples of Cultural and Sociopolitical Questions about GMOs

1. Regulatory Issues. Do we have regulatory and biosafety mechanisms to make sure that citizens are protected?
2. Risks and Benefits. Are people concerned about the distribution of risks and benefits among consumers, farmers, corporations, and others?
3. International Trade. Should we invest in a technology that cannot be exported to some countries?
4. Consumer Choice. Is the labelling debate about consumers' having the right to choose what they are eating?
5. Effects on Rural and Developing Communities. What will genetic modification technology mean for small-scale farmers?
6. Nature Tampering. Do we have the right to alter things that God has created in nature?

Source: Based on Brossard, workshop presentation, slide 4.

genetic engineering on rural communities are raised. Other cultural groups raise concerns about tampering with nature. In other words, the meaning of a new technology can be multidimensional, and the issues extend beyond questions about the science into social, ethical, and legal questions.

Debate among people who have varied interests related to those complex issues is healthy in democratic societies, Brossard stated. However, debate is situated in particular social, cultural, and political climates at a given time. Strategies for engaging the public in discussion about scientific issues should take the differences into consideration, she suggested. Because societal discussions about GMOs are so multifaceted and complex, "there is potential for polarization," she concluded.

THE PATH TO POLITICAL POLARIZATION

Roger Pielke of the University of Colorado clarified three related concepts: policy, politics, and politicization of science (Box 4-2). "If there is no choice to be made and there is no decision to be made, you are not engaged in policy," he stated. He also noted that policies are not just a government function but that universities, businesses, and other institutions have policies. Although *politics* has come to have a pejorative tone, Pielke explained, by definition it is simply the way in which the business of living together in society is accomplished. Thus, when science, policy, and politics are combined, you get the politicization of science. If the role of science is viewed in that light, science should be tightly integrated with politics because it can serve a useful function in helping people to make better decisions, but the pathological politicization of science (intentional politicization for the purpose of personal gain) is to be avoided, Pielke argued.

BOX 4-2
Definitions of Science, Policy, Politics, and Politicization of Science

Roger Pielke provided lay definitions to frame discussions about the intersections of science, policy, and politics.

- Science: The systematic pursuit of knowledge.
- Policy: A decision.
- Politics: Bargaining, negotiation, or compromise in the pursuit of a desired end.
- Politicization of Science: The use of the systematic pursuit of knowledge as a means of bargaining, negotiating, and compromise in pursuit of a desired end.

Source: Pielke, workshop presentation, slide 8.

How does science become polarized? Dan Kahan of Yale University addressed that question by discussing what he termed "the science communication problem, the failure of compelling widely available evidence on risk and related facts to quiet dispute about what those facts are even when the evidence directly speaks to it." Kahan refuted the idea that the science-communication problem can be attributed to public science illiteracy, public distrust of science, or orchestrated misinformation campaigns (See Chapter 2). Rather, the cause of the science-communication problem is a "polluted science-communication environment" in which there is widespread disagreement along political lines about facts, and the disagreement is exacerbated by motivated reasoning and confirmation biases, Kahan said. People seek affiliation with others who are like them, and groups on both ends of the political spectrum have people who are science-literate and have effective mechanisms for conveying what they know to others. Pielke noted that "polluted science-communication environment" is another way of saying "politics". Kahan listed climate change, private gun ownership, and fracking as examples of highly politicized, and hence polarized, societal issues (a polluted communication environment). He stressed that this degree of polarization around a scientific issues is not normal.

Issues that divide people into political camps are called wedge issues, Pielke said. According to him, science is increasingly seen as a potential wedge issue in modern politics, with more academic scientists and experts participating in the process than ever before. He argued that that has happened in part "because science gets greater standing when it is politicized. Academics get greater visibility, and there are political gains even if it does affect the science-communication environment pathologically."

Fortunately, Kahan emphasized, there are far fewer science issues in a polluted communication environment than in an unpolluted communication environment. Widespread polarization "happens when issues of fact or risk that admitted scientific investigation become entangled in social meanings that transform positions on them into badges of membership, at which point people will have more at stake in fitting in with their group than only forming a position that is convergent with science." In a polluted science-communication environment, people are told not only the facts but *who* believes *what*, he explained. That turns matters of science into "us vs them" situations, Kahan concluded.

It is possible to create polarization where it does not yet exist. Kahan used childhood vaccination to illustrate the point. According to the Centers for Disease Control and Prevention, vaccine rates are high and have remained so: less than 1% of children receive no vaccines, Kahan said. He explained that vaccination is generally viewed as a public good with people contributing in a reciprocal fashion. If people believe that others are contributing, they are happy to do so also; however, it is risky to make people underestimate the degree to which other people are contributing. On a host of issues, Kahan continued, Americans are divided; however, there is little political division about vaccines. As he explained, it would not be difficult to create a polluted science-communication environment; "all you have to do is create the conditions in which people are going to start to think, 'I did not recognize that that was one of the positions on which it is us vs them.' " Effective science communication is using the information that we have about how people come to know what they know to make sure that we get the benefit of all we know as the result of science, Kahan said.

Genetically modified foods do not fall into the category of a polluted science-communication environment, Kahan said. He explained that the science of GMOs is not being debated by members of the public; they have no opinion, they know little about them, and most people still consume them. That is why polls on this topic do not reflect how people will vote on the labeling of GMOs, he said. If GMOs are debated and become the subject of polls and referenda, it is due mostly to the efforts of particular interest groups, not because of public opinion about the science, he suggested. Science-communication research has a role to play in identifying the source of the problem, which may not be the way in which people are processing information, he added. He argued for using what has been learned through social science to keep GMOs from becoming a polarized topic.

MODES OF SCIENCE ENGAGEMENT WITH PUBLICS AND POLICY-MAKERS

Pielke discussed the role that scientists can and should play in the political environment. He described how the role of science in policy-making was recently at the forefront in the discussion about GMOs at the European Union. In 2013, Ann Glover, the chief science adviser (CSA) for the European Commission, publicly suggested that science does not support assertions that GMOs are dangerous. José Manuel Barroso, president of the European Commission, responding to a query from a member of the European Union parliament as to whether he agreed with Glover's comments on GMOs, stated that "the CSA has a purely advisory function and no role in defining Commission policies. Therefore, her views do not necessarily represent the views of the Commission." Ultimately, Glover was removed from her role, and the science-advisory structure in the form of a CSA was eliminated. A key problem that Glover faced, said Pielke, was that her role as a science adviser had not been formalized, prescribed, or well understood.

Pielke emphasized that all communication and engagement is political if it concerns what ought to be done about an issue. To help scientists to navigate this terrain, he developed a set of guidelines for experts who have to engage with the public and policy-makers; it was published in his book, *The Honest Broker.*[20] The work details different modes of engagement and draws four main conclusions about roles and responsibilities when scientists engage with decision-makers and the public:

- Discussing roles and responsibilities is important when scientists engage with decision-makers and the public.
- Scientists can play multiple roles, all of which are important.
- All communication by scientists in the public realm is political, despite the desire of scientists to simply inform, elevate the discussion, or stay removed from the political process.
- Institutions play a critical role in public engagement between science, members of the public, and policy-makers.

[20]Pielke, Jr., R. A. 2007. *The Honest Broker*. New York, NY: Cambridge University Press.

In his book, Pielke describes four idealized modes of engagement for scientists and experts: the pure scientist, the issue advocate, the science arbiter, and the honest broker. The four types are based on how people believe a democracy should function and on their views on the role of science in society. Pielke described each of the four types and the implications of each, using the example of providing guidance on which restaurant to choose for dinner in Washington, DC.

The spectrum of possible roles spans from advocating for a specific restaurant (go to McDonald's) to laying out all the options (the yellow pages or a restaurant travel guide), Pielke explained. The pure scientist might say, "I don't want anything to do with your values-based decisions about food. Fortunately, the US Department of Agriculture [USDA] has these dietary guidelines about what to eat for dinner. This will empower you to make decisions about what to eat for dinner," Pielke said. However, he added that even those dietary guidelines are not as "pure" as they might seem. The USDA dietary guidelines include meat, but "you do not have to have meat for a healthy diet." The information that the pure scientist provides will always involve the choices and motivations of actors who have stakes in the choices. Pielke argued that once a scientist engages with the public or with policy-makers, he or she has stepped out of the role of a pure scientist.

The science arbiter is like the concierge in a hotel. The concierge is able to answer empirical questions, such as, could you tell me the three closest Italian restaurants? The science arbiter is the expert who answers questions, but the person making the decision drives the conversation. Members of Congress ask scientific questions of panels of experts, who report to them. That mode of engagement has been criticized with questions about who serves on such expert panels and how they are selected.[21] The key in the case of those panels is that there is an informal engagement between the decision-maker and the expert around an issue, in contrast with the case of Ann Glover, who was not asked to assess GMOs, Pielke explained. The lack of an institutionalized mechanism for soliciting advice was problematic in her case, he argued.

The defining characteristic of issue advocates is the role that they play in narrowing the scope of choices of the decision-maker, Pielke said: "You may tell me to go to dinner at McDonald's." He stressed that advocacy has a long history dating back to the Federalist papers and is fundamental to American democracy. "People who tell you that we should label genetically modified foods are advocates. People who tell you that we should not are advocates." Science often gets enlisted in advocacy campaigns because scientists enjoy high standing among many members of the public.

The honest broker is like a travel guide, Pielke said, offering a variety of choices without making a specific suggestion. In the restaurant example, the honest broker provides "the yellow pages of all the restaurants in the Washington, DC, area". Honest brokers do not tell you what neighborhood has the best restaurants or which restaurant to choose; "they tell you what your options are."

Some have argued that a science communicator that does not want to be involved in politics exemplifies a fifth role in which scientists can engage with policy-makers. However, Pielke explained that that role does not exist. Instead, he said, "what happens is that we pretend that we are pure scientists or merely science arbiters. We are just talking about the facts. But, in reality, what is going on is an effort to use science to try to motivate a particular set of decisions or often a particular decision." When such *stealth advocacy* happens, especially in a polluted science-communication environment, people become wary of the motivations of the communicator. The problem is especially prevalent in conversations about climate change, Pielke stated.

Figure 4-1 is a flow chart that describes when each science role might be appropriate; each role has value that depends on context. Each expert must decide what role to play on the basis of the degree of values, consensus on the issue, and the presence of uncertainty. Pielke added that it is virtually impossible for a person to play the role of an honest broker. The honest broker would best be a "diverse committee of experts in authoritative, legitimized institutions, such as the National Academy of Sciences", he said. He cautioned that such organizations as the National Academy of Sciences, the Royal Society, and the Intergovernmental Panel on Climate Change threaten their legitimacy in the

[21]Jasanoff, S. 1990. *The Fifth Branch: Science Advisors as Policymakers*. Cambridge, MA: Harvard University Press.

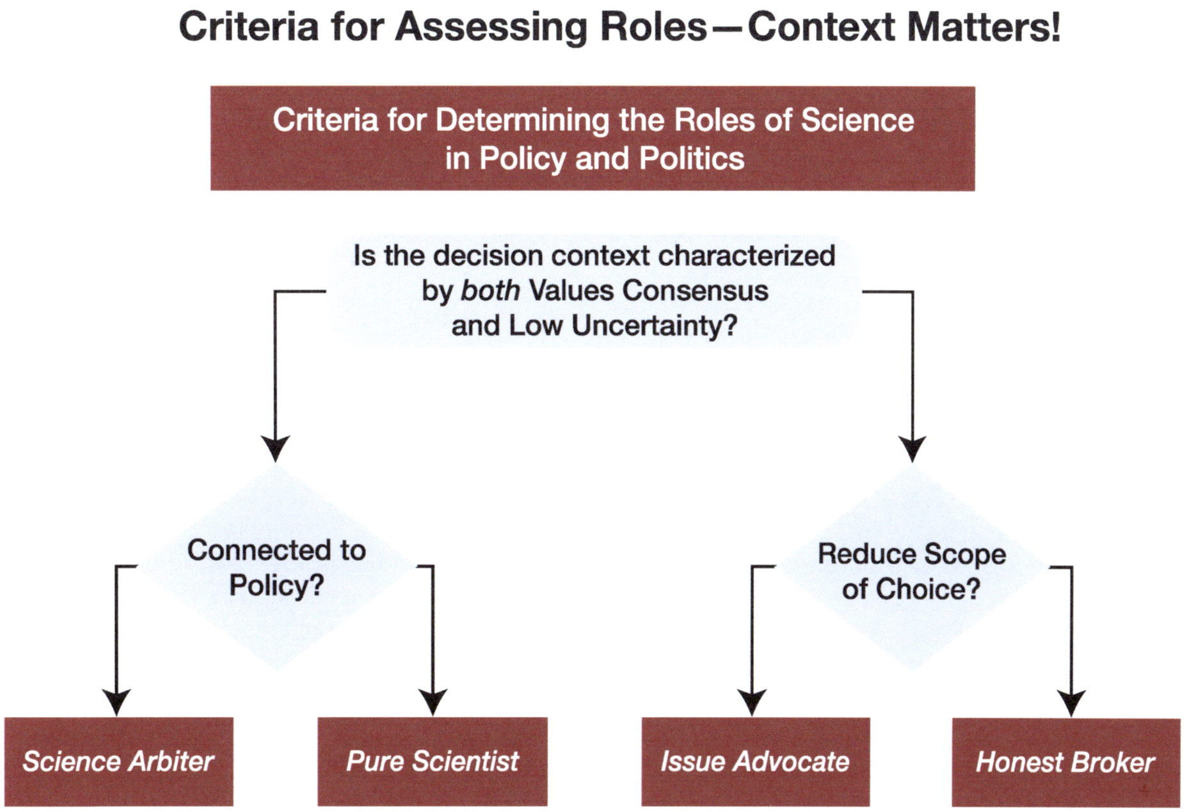

Figure 4-1 Criteria for assessing science communication roles in policy and politicalpolitic contexts matter. SOURCE: Pielke workshop presentation, slide 12.

honest-broker role if they are seen as engaging in advocacy. Such highly respected institutions are rare, so loss of legitimacy can be especially important.

HOW SCIENCE INTERSECTS WITH POLICY

Pielke described several types of science issues. For some issues, information does not matter; it is used only as a symbol or to support existing positions. For these issues, such as in debates about abortion, there is a lack of agreement about the ends or unmanageable uncertainties. Pielke calls that situation *abortion politics*. In contrast, *tornado politics* involves issues on which a decision has to be made (a tornado is approaching), people agree on the ends or means (such as staying safe in a shelter), and there are manageable uncertainties. Tornado politics is considered to involve *tame problems* in that they are relatively easy to solve with data, facts, or evidence. *Wicked problems* are ones that entail tremendous uncertainty about the variety and effectiveness of solutions and highly conflicting societal values. How the world should address climate change is an example of a wicked problem.

When people treat wicked problems as though they are tame problems, two effects can occur. First, scientists often rely on the deficit-model approach to make a policy decision, Pielke said. For example, "if only people understood the science of climate change, we would all agree." Second, reframing of wicked problems as tame problems often leads to stealth advocacy. With wicked problems, it is easier to use science as a wedge issue and pollute the science-communication and decision-making environment.

Pielke acknowledged that those roles and types of issues can be frustrating to scientists who want to engage in science-based discussions but avoid politics. However, trying to use science to tame a wicked problem often has the effect of worsening the science-communication environment because the science does not address the conflicts in societal values. Pielke argued that the existence of wicked issues underscore the importance of having institutions with diverse panels of experts to present legitimized views of the state of science.

David Goldston of the Natural Resources Defense Council offered his reflections on how science intersects with public policy. He identified four types of policy and science intersections: a policy question masquerading as a science question, a science question from policy-makers when broad consensus exists in the scientific community, a science question from policy-makers when there is little scientific consensus, and a question about a science issue on which policy positions and the science are undecided. Those intersections occur because "the goal of everybody in a policy debate on all sides is to say that it is a question of science, because if you can say that science is on your side and convince people of that, you win," Goldston stated. That is effective because scientists are highly regarded. When debates are highly polarized, both sides claim that the facts support their position and that those who oppose their position are wrong.

Policy issues masquerade as science issues when the science is essentially decided and a decision must be reached. For example, for some air pollutants, the science has been clear that particular degrees of air pollution are associated with particular numbers of hospitalizations, Goldston said. Therefore, the debate centers around the target for the pollution magnitude or around how many hospitalizations are acceptable. Politicians are loath to discuss the policy and its implications and instead focus on the scientific data on the pollutants and engage scientific advisers. Ultimately, in this case, the debate became very heated and led to sides that were polarized on the basis of their views on health policy that still exist, Goldston explained.

The second type of policy–science intersection occurs when a science question from policy-makers is asked and there is broad consensus about the issue in the science community. Issues around climate change fall into this category, according to Goldston. It differs from the first category because policy-makers are debating a true science question rather than debating a policy choice that they have to make. Climate change is a high-profile example, but this situation is relatively rare.

Far more often, policy-makers ask science questions about which science has not reached consensus—the third type of intersection of policy and science. For example, when policy-makers ask science questions about the effects of GMOs on ecology, they may receive a variety of answers from the scientific community. The ability of science to answer policy questions often lags behind the timeline for making decisions. Ultimately, deciding what to do in the face of uncertainty is a policy question, Goldston explained, and the decisions become value questions. Policy-makers often use scientific uncertainty as a distraction when policy decisions are difficult, he added.

Finally, some science questions or technologies are so new that both the policy positions and the science are unsettled. In such cases, what all sides and stakeholders want is more science. "You can tell when an issue in Washington is not fully mature yet, because the debate is less immature," Goldston joked. The environmental consequences of nanotechnology constitute an example of such a topic. However, as the science advances and people adopt policy positions, the research can be called into question. Policy-makers may mistrust the data or call into question the motivations behind the funding agency or researchers involved with producing the data.

In light of those four broad types of science–policy intersections, Goldston moderated a panel discussion on the role of scientists in public policy decisions on whether to label genetically modified foods.

THE ROLE OF SCIENCE AND SCIENTISTS IN SOCIETAL CONVERSATIONS ABOUT LABELING OF GENETICALLY MODIFIED FOODS

Goldston asked a five-member panel to consider what type of intersection the labeling of GMOs represents and what the roles of scientists are in this debate. The members of the panel were Robert Goldberg, professor of molecular, cell and developmental biology at the University of California at Los Angeles; William Hallman, professor and Chair of the Department of Human Ecology at Rutgers University; Tamar Haspel, food and science journalist with the *Washington Post*; Eric Sachs, environmental, social and economic platform lead at the Monsanto Company; and,

Allison Snow, professor of evolution, ecology and organismal biology at Ohio State University. The following subsections present the topics considered by the panel.

Is Labeling of Genetically Modified Organisms a Science Question?

Goldston asked the panel to consider whether the debate is about the public's right to know whether GMOs are in the foods that they eat and what role, if any, scientists and other experts have, beyond their participation as citizens, in the debate over whether to label GMOs.

Hallman pointed out that the roles of social scientists and natural scientists in policy debates may differ. From a natural-science perspective, great challenges exist in determining useful thresholds for a labeling regime. Noting the 0.09% threshold for allowable GMOs in foods in the European Union, exceptions for particular ingredients, and varied laws, he questioned whether such regulations had any scientific basis or could be answered by scientists. Social scientists have a role to play in helping decision-makers to understand what labels would trigger in the minds of consumers on the basis of their research. Available data show that labels give consumers impressions that may not be scientifically true, Hallman said. For example, using a threshold approach frames GMOs as posing a problem when its concentration is above some threshold and not a problem when it is not. Thus, policy-makers would need to weigh whether requiring labels that trigger false impressions qualifies as mislabeling.

Haspel responded to Goldston's query with the view that GMO labeling is not a question of science but a question of utility. Food labels in general are not guided by any "grand unifying theory" that helps people to discern what should be on labels. She argued that it is unclear why some components are labeled and others are not. "We label vitamin A, but not vitamin D. Why do we pick the things that we pick and not pick the other things?" Thus, labeling is about more than science, and neither science nor policy can provide clear answers on what belongs on a food label and what does not. Haspel asked participants to consider that a vocal minority has strong anti-GMO feelings at the same time that people have strong affinities for particular foods that contain GMOs. "If we were to label, we would force people to choose: they could no longer have both their grievance and their Doritos." She mused about whether forcing people to choose between their ideals and the food they use to feed their family would make societal debates about GMOs go away. Goldston took that notion a step further, asking the panel to consider whether people's concern over ecologic effects of GMOs could cause them to oppose other people's eating of foods that contain GMOs and ultimately change the role of scientists in this debate.

Sachs noted that people who oppose GMOs do so for a variety of reasons, so "the question that I always ask in these conversations is, What will be the outcome if we have labeling?" Sachs also indicated that, on the basis of what happened in Europe, a likely consequence of labeling would be a reduction in food choices. In Europe, food manufacturers and grocery-store marketers do not provide their products with and without GMOs. Instead, they followed consumer demand—a realm that does not directly involve scientists. He stated that he would like to engage in conversation, understand concerns, and provide evidence and education to people to help them to reframe their positions.

Goldberg described the role that he played in crafting the arguments against GMO labeling in California's Proposition 37.[22] He argued against mandatory labeling because in his view it reflected poor policy. Presenting the science about exactly what genetic modification is was part of explaining the arguments against labeling. Goldberg sees engaging in discussion about GMOs as a challenging but necessary part of ensuring that policy about GMOs is rational and based on science. However, he contended that his role as a scientist is to provide people with the facts about genetic engineering and not to make a policy decision.

Snow indicated that ultimately labeling GMOs in foods is not a scientific question. "I think that scientists can provide a lot of good information that people might have questions about when they are making decisions." In other words, science could inform people's opinions, and she would not answer

---

[22]See http://vig.cdn.sos.ca.gov/2012/general/pdf/37-title-summ-analysis.pdf for more detailed information about California's Proposition 37.

questions beyond the science.

Sachs explained further that he thinks that the role of scientists "is to help people to identify the various kinds of consequences of their decision-making process." He added that he tries to make rational decisions by using the information available. In his opinion, the benefits of GMOs are not widely known and are not being considered by most when they are reaching personal or policy decisions about GMOs. Hallman countered that scientists do not need to be in a situation where they are seen as "an authoritative parent saying 'You should eat your peas because they are perfectly safe and they are good for you.' We have to have more of a conversation than that." Yet, noted Goldston, the safety of genetically modified food is not merely a matter of opinion, as is taste or texture.

Intended and Unintended Consequences of Labeling

Goldston framed the conversation about the consequences of labeling as a thought experiment. He asked the panel to consider the idea that proponents of GMO labeling want the labels to be interpreted as a warning to consumers. He then asked the panel to respond about whether it is legitimate to have a warning for genetically modified foods or foods that contain genetically modified ingredients and about the role of the scientist in such a debate.

Haspel noted that determining whether labeling foods that contain GMOs is beneficial or harmful depends on a persons' assessment of the likely consequences. However, with little certainty about what farmers and consumers will do, predicting the effect of a policy choice is impossible. Thus, predictions of the societal impact of GMO labels are dependent on the worldview of the predictor, she stated.

Hallman reminded the participants of ecolacy, the ability to predict the intended and unintended consequences of particular actions. There are relatively few GMO products, but if labeling is mandated it will establish the regulation of all later GMO products. Hallman argued that there is too little information about the technology itself and about its benefits and drawbacks. For example, opposing crops that are "Roundup Ready" because one dislikes pesticides may yield one decision, whereas learning that "Roundup Ready" crops may reduce the overall amount of pesticides used may yield the opposite decision. In either case, a label that says "contains or may contain GMOs" does not provide enough information for the decision and may have the effect of barring crops that would have particular health benefits.

Goldberg stressed that the issue is not labeling vs no labeling, inasmuch as food labeling already exists and can be instituted by companies at their discretion. He said that the societal discussion is about whether labeling should be mandatory. Mandatory labeling leads to the perception that something is harmful or negative, he said. Second, he emphasized that public policy does not occur through election or opinion. In his view, mandatory labeling would lead to fewer choices. Therefore, perhaps voluntary labeling is the middle ground that provides the public with information but does not involve an arbitrary threshold. When asked whether voluntary labeling would result in greater confusion and opportunities for misleading, Goldberg pointed to the lack of regulation of the natural-food industry, which offers dietary supplements with no Food and Drug Administration oversight.

Sachs pointed out that reducing choices and having fewer items on grocery shelves have other upstream consequences. For example, farmers will plant fewer genetically modified crops, so whatever benefits farmers and the environment might have received from using them will be eliminated. That could result in a return to previous farming methods that used more chemicals and less conservation tillage. Thus, a question about labeling has implications beyond what products and choices are available.

Scheufele asked the panel to comment on what has been learned about consequences, motivations, and benefits from the voluntary labeling of GMOs in the United States. Hallman answered that most US citizens are uninformed about labeling. Research suggests that people assume that labeling is already mandated and that the organic-food standards include being GMO-free. Hallman added that the GMO-free label has more effects on particular products. For example, people seeking more wholesome foods might weigh a GMO-free label more heavily than someone seeking a processed food from a convenience store. He suggested that there may also be differences between particular brands of a single company.

How Much Regulation Is Enough, and What Is the Role of the Scientist in Determining That?

Regulations determine both whether labels are mandatory and what type of information they will include. Goldston asked the panel to consider what they believe to be the appropriate amount of regulation and how science informs that Snow answered in brief that it is not sufficient to have no regulation. Sachs stated that some regulation is appropriate, but expressed concern that excessive regulation might limit the progress and application of modern processes. He asked "is it true that more regulation actually leads to greater comfort and acceptance that something is ok and safe?" Goldston countered that social scientists, historians, and economists have demonstrated that regulation "has made a huge difference in terms of making the public feel safe." Hallman and Haspel both indicated that they did not know how natural or social scientists can contribute to discussions about how much regulation of GMOs is appropriate. Goldston remarked that sometimes saying that you don't have an answer "is the most important answer."

5

How Should Scientists Engage in Conversations about Genetically Modified Organisms?

"Where there is respectful discourse, it is my experience that we get better-quality outcomes in the public interest."—Molly Jahn

This workshop focused on how people process information, how context shapes how people view and use science, and the intersection of politics, policy, and science. The case of GMOs illustrated how those factors come into play around a specific science issue in the public sphere. On the final day of the workshop, participants engaged in breakout sessions, and plenary discussions about scientists' practical application of the lessons of the science of science communication.

CONCEPTUAL TAKE-HOMES

To set the stage for a discussion of conceptual and practical take-homes, Brooke Smith, executive director of COMPASS, summarized key conceptual points made by speakers and panelists during the course of the workshop (Box 5-1). As a self-described optimist, Smith noted that although participants had expressed feeling depressed about how little facts appear to matter in how people make decisions, she had come to believe that "science matters a lot and facts matter a lot, and there is a big role for science and scientists in society." She reiterated a comment made by David Goldston of NRDC that scientists should not overcorrect for the deficit model and conclude that facts do not matter at all.

The workshop demonstrated that "we know a lot about the science of science communication, how people make decisions, how opinions are formed," Smith said. "Now that we know those two things, what does that mean that we should do?" Smith reminded workshop participants that the principles of how people process information, make decisions, and engage with the world apply to everyone—including scientists. She urged participants to keep those principles in mind as they consider the specific cases of public interfaces with GMOs in breakout group discussions. The breakout groups focused on three specific cases of GMOs: the reintroduction of the American chestnut, the Oxitec mosquito abatement program, and the issues surrounding Bt (transgenic) corn and the monarch butterfly. The plenary reports—which included the science claims, societal implications, and public-engagement considerations—of the breakout group discussions are described in Boxes 5-3, 5-4, and 5-5 at the end of this chapter.

PRACTICALTAKE-HOMES: WHAT CAN SCIENTISTS DO?

Over the course of both days of the workshop, the presenters and participants shared their thoughts on practical take-homes that may help life scientists and their supporting institutions to prepare for and conduct public engagement in science. The ideas emerged from presentations and panel discussions during the first day of the workshop, from the reports from the breakout groups on the second day, from audience discussion, and from a final response panel that comprised of Rick Borchelt of the Department of Energy, Helene Dillard of the University of California at Davis, Molly Jahn of the

BOX 5-1
Key Points from Workshop Presenters and Discussants

Brooke Smith synthesized the key points that she took from the first day of workshop presentations and discussions:

- The public trusts scientists but might not listen to them when scientists' information conflicts their own worldviews or values.
- The deficit model is the idea that if people had more information, they would agree with scientists and make better decisions. But that is not how people make decisions. Better explanation by scientists and better listening by the public is not the answer to the science-communication problem.
- Many decisions are based on emotions more than on cognitive processing or reasoning.
- People use mental shortcuts to make sense of excessive and complex information.
- All people engage in motivated reasoning, the tendency to seek information from others like them, which reinforces their own views.
- People are subject to confirmation bias, the tendency to look for information that supports what they already believe and to dismiss disconfirming information.
- Bringing people with diverse viewpoints together is one way to resist confirmation bias.
- The context of science communication includes multiple cultures and values.
- All information is framed, regardless of the intent to frame.
- Some believe that the GMO issue has already been negatively framed, and others see GMOs existing in an unpolluted science-communication environment because so many people are not engaged on the topic.
- Scientists can assume various roles in a policy discussion—pure scientist, science arbiter, issue advocate, and honest broker—and all the roles are needed in a robust society.
- The pure science communicator, uninvolved in politics, does not exist.
- There are different types of problems—tame problems and wicked problems. Communication needs to be suited to the problem, especially when there are conflicting values.
- Closer examination of the case of GMO labeling indicated that one of the roles of natural scientists involves asking ecological questions that are not being asked and that one of the roles of social scientists is to help people to understand what labeling decisions would trigger in consumer behavior.
- Everyone has a role in a public policy debate, including scientists, although GMO labeling may ultimately not be a scientific question.

University of Wisconsin-Madison, and Dan Kahan of Yale University. The take-away ideas—from improving translation of social-science research to wrestling with personal biases—are described below.

Understand the Nature of the Controversy

Kahan emphasized that the presence of small groups of passionate stakeholders does not suggest that the wider public is in a state of division about GMOs. He cautioned participants about generalizing from their own experiences, noting that the reason that many workshop participants are involved in controversies about GMOs is that they "live inside an environment in which everyone actually is a stakeholder." He argued that social marketing campaigns to address controversies about the technology would not reflect the values of the larger public and that ensuring that people get all the information on the science issues around GMOs would be the wrong science-communication goal. Taking the time to study and identify the true issues to develop appropriate strategies is a better approach, according to Kahan. He challenged participants to consider how they could foster a science-communication environment that supports basing decisions on diverse values and interests while gaining access to the best information available. He noted that people have to recognize and act on more information than they can comprehend. Most people get information through other people in their communities who have an interest in a topic and show that to others through their attitudes, words, and actions. Kahan explained that some of the worst problems caused by miscommunication in science occur because people are being prevented from having a reliable view of the cues that they use to recognize what is known in science. He emphasized that engaging key stakeholders and ensuring that they know that their concerns are being taken seriously is an important step because others in the community often take their cues from them.

Scheufele challenged the notion that GMOs exist in an unpolluted science-communication environment. In some settings, the opposition to GMOs is intense and includes regular threats to people and property, he said. But in agreement with Kahan, Scheufele urged participants to resist the temptation to believe that simply finding a way for people to get more reliable information—a return to the deficit model—will solve the science-communication problems surrounding GMOs.

Manage the Communication Process

Kahan noted that discussion in his breakout session focused on the idea that "if you do a good job in managing the science-communication environment, and in particular the stakeholder settings, if you don't just try to impose something on people, if you enter into their lives in a way that shows that you are respectful of and solicitous about their stake in what you are doing, you will have a community that is less likely to be vulnerable to being misinformed."

To illustrate the point about needing a communication plan, Dillard pointed out that a lot of her communication as a dean of a school of agriculture is about damage control. She added that there are always two sides to the issues that she must address: "one side is happy, and the other side is angry." Borchelt emphasized that managing the communication process around GMOs can follow only from understanding what the problem is. The science community lacks an overall "game plan" for managing communication about GMOs, he added.

Draw from Social-Science Evidence to Inform Public Engagement Practice

Multiple participants emphasized that scientists who communicate their work and engage with the public could benefit from learning from social-science research that helps to explain how people process information and to explain the effectiveness of different approaches to communication. Trevor Butterworth of Sense about Science remarked that making use of social-science research can keep scientists and science communicators from "reinventing the wheel". Kahan echoed the need for communication about GMOs and other science topics to begin with evidence about how people come to know what they know and with the context of communication. He concluded by saying that "our common enemy is that we might not get the benefit of the common knowledge. It does not matter what our positions are, because whatever our values are, we will not be able to achieve what we want to achieve if we do not understand all the best evidence that we have."

Butterworth emphasized that creating a repository to manage, house, and share knowledge about best and worst science-communication practices is critical. Such a repository, perhaps maintained by an academic center or a national science organization, would be vital in ensuring that public engagement about potentially controversial science issues is informed by the lessons of social science from the outset, not as an afterthought, he added. Bruce Lewenstein, of Cornell University, noted that institutions can also devote the needed time and resources to monitor the science-communication landscape that scientists themselves do not have.

Learning from the experience of those who are in the field and are communicating with the public may be an important step. Kahan said that companies, such as Oxitec, that are engaging the public or conducting surveys about science should record, study, and share their experience in communication. Doing so could help to reduce future missteps for themselves and others.

Get Communication Training

Training is an important step that scientists can take to be prepared for engaging with the mass media, explained. "If you are publishing in a journal like *Nature*, which a lot of advocacy groups are going to be paying attention to, you have to be ready to look at the big picture when you are communicating about your research," Snow observed. She emphasized that scientists need to think carefully about where they will publish, whether to talk to the press, and what types of reactions people might have to their work. Preparation and planning in a university setting should be coordinated with the university's press office. Often, the press office has a greater interest in generating hype, although some are more conservative. Having a strategic plan in place for handling inquiries can be advantageous. As Snow stated, "we cannot control how these stories are going to unfold, but we can be aware of the possibilities and try to prepare to get good information out there from all of different people who are providing the information."

Examine Your Biases

Tamar Haspel a food and science journalist discussed the importance of examining your biases. Haspel pointed out that most people want to parse data accurately, understand the issues, and communicate persuasively. How can that be done? Haspel affirmed that pushing back against the human tendency to seek confirming information (confirmation bias) is especially important in her role as a journalist. Haspel has devoted much effort to understanding her own biases and has developed a series of steps to improve how she and others communicates about science (Box 5-2).

We must be convinced that we, ourselves, are subject to motivated reasoning, Haspel said. It is tempting to believe that only other people are subject to motivated reasoning, but in fact all people do it. "We have to wrangle our own elephants before we can start thinking about other peoples' elephants," Haspel stated. She suggested two exercises that can help people to recognize and address their biases:

- Identify a position that you are wrong about. Haspel noted that one's own position always seems right and it is easier to see the inconsistencies in the positions of others. She emphasized that it is challenging for people to get used to the idea of being wrong in some of their positions.
- Think about instances in which you have changed your mind on a substantive issue. Haspel remarked that it can take a long time for people to change their minds. "The idea that we go out and expect people's minds to change in an evening or in a discussion is unrealistic," she said.

"We must reconsider bias" Haspel said. She stressed that bias "is not something that is bad," but we must be able to negotiate between our biases and those of others. Haspel remarked that it is important to vet your sources of information and manage your media. There is a strong temptation to believe that sources are credible when they agree with one's beliefs. Haspel's test to determine credibility is to find a source that will acknowledge both risks and benefits associated with an issue. "In any kind of complex issue like genetic modification, there are going to be pluses and minuses, there are going to be compelling arguments on both sides." Sources that note only one side of an argument may not be wrong, but they probably have "an ideological dog in the fight", Haspel stated. She also looks inward and tries to identify and acknowledge both the risks and benefits associated with science and technology. To be exposed to multiple viewpoints about those risks and benefits, Haspel suggested that people manage their media. "Are there people on both sides of the GMO debate on your Twitter stream? Make sure that there are," she said. She also stated that in addition to diversifying her media sources, "I try to find the smartest people I know who disagree with me and call them and ask them questions, and then I listen." Haspel explained that although that practice allows her to learn the best argument for the opposing position, it also causes her to change, temper her own opinions, and become more open-minded. She pointed out that it is good practice to talk not just about scientific facts but about the

BOX 5-2
Small Steps Toward Better Communication

1. Be convinced. Elephant wrangling begins at home.
2. Reconsider bias.
3. Vet your sources, and manage your media.
4. Acknowledge both risks and benefits.
5. Find the smartest person who disagrees with you, and listen.
6. Understand and appeal to values. Tell stories about people.
7. Reach across the aisle.

Source: Haspel, workshop presentation, slide 23.

stories and values of the people involved with science. She commented that citrus-greening story written by Amy Harmon[23] is so good because "it told the story not just of a plant but of the people whose livings depend on the plants and how they are struggling with the issues." However, the most important thing that people can do is talk to one another. "I am pretty convinced that the key to peace in our time on GMOs is getting people with various points of view together in a room." She concluded that recognizing that bias is part of the human condition rather than a shortcoming can advance how we engage with others about science in a rational way.

Share the Human Story

Dillard noted that ensuring that there is a human side to the story is important if we are to avoid allowing people to be portrayed as villains, as is acknowledging when new technologies have consequences that need to be addressed or mitigated. Lewenstein added that attaining the right mix of storytelling and evidence about a topic is an approach that science communicators have been discussing.

Sometimes No News is Good News

Borchelt pointed out that science does not have a unique pull on people's attention. He added that scientists need to take into consideration that a particular finding or subject of research will become salient to people at different times, depending on personal or societal contexts. However, "at the end of the day, few of these conversations are national conversations that are going to affect any national political outcome," he stated. In a similar vein, Jahn pointed out that a wide array of regulated technologies are relevant to food systems and that "the vast majority of them no one ever hears about." She said that regulatory processes are "critically important for negotiating the public discourse about the fate of a technology and its implementation in food systems" and that often, "no news is good news."

[23] Harmon, A. "A Race to Save the Orange by Altering Its DNA." July 27, 2013. New York Times.

Diversify the People in the Room

Some workshop participants emphasized that it is critical for multiple viewpoints to be represented. Engagement and discussion about science issues occur in a large framework in which scientists do not have the only voice, Scheufele stated. Borchelt noted that there is no consolidated "anti-GMO" message or group, but rather multiple publics with various interests. Tim Schwab, of Food & Water Watch, commented that "it is worthwhile to respect and include a diversity of scientific opinions" in both conversations about GMOs and events like the present workshop that focus on public engagement about GMOs.

Jahn stated that, particularly for those who have been immersed in the controversy surrounding genetic engineering and foods, the conversation has been idiosyncratic and expensive. She observed that early in the discussions about genetic engineering, insufficient attention was paid to the different values that people place on innovation. "The science and technology community tends to think that innovation equals great, but that is not how everyone feels. The science community has had a huge blind spot, and probably also science communication, on this particular issue." She added that communication about the unintended and foreseeable consequences of innovations should be improved, and she called on her colleagues to avoid dismissing and discrediting or otherwise failing to respect legitimate concerns about the consequences of new technologies.

Public Engagement Is Not the Same as Persuasion

Brossard emphasized that communication has a role to play in limiting polarization and controversy. She challenged participants not to shy away from communication "that promotes meaningful public engagement." Borchelt and Jason Delborne of North Carolina State University pointed out that engagement between scientists and citizens is not synonymous with persuasion. "You need to be able to put yourself at intellectual risk to have true engagement with other parties," Borchelt commented. According to Haspel, distinguishing engagement from persuasion involves ensuring that people who have diverse interests are truly represented in a conversation and making an effort to find common ground.

Jahn noted that one of the highest costs in how GMOs have been debated is that the scientists who have been involved for many years cannot have open-minded two-way conversations, because they have been injured by the roles that they have played over the years. However, everyone cares about food, Jahn noted, and more nuanced, complex conversations may be more feasible for younger scientists.

At times, scientists and academicians are approached by members of the public who have not formed an opinion and want information about a new technology, such as GMOs, Dillard stated. "What do you want us to use if we do not fall back on the data?" she asked. Delborne suggested that such questions from the public form the entry point for a conversation, rather than only an invitation to provide information. Such a conversation could begin with returning the question and asking them what they know and care about with regard to this issue before offering to share their own information and cares. Another way to consider the questions that people have about a new technology, Brossard suggested, is to keep in mind that concerns may not be limited to health and environmental concerns but may also include ethics, corporate monopolies, or the right to know. Those concerns need to be represented in important conversations.

Common Ground Encourages Respectful Discourse

A number of presenters and workshop participants emphasized that high-quality discourse should be at the heart of public engagement about conversations about food-related technologies, the science issues. Although a number of public institutions have effectively brokered productive existing regulatory system has failed in that regard, Jahn said, particularly with regard to addressing the vested financial interests of various parties. Yet, in her view, when the discourse is respectful, better-quality outcomes in the public interest emerge. "Here is a message I have tested many times in many frames: Everybody knows that food is important. That is well beyond the fact that it is our source of survival, our source of nutrition every day. No one wants to wreck the planet by growing food. Diversity is the sign of a healthy system. " No one argues with those points, and they can be used to frame many discussions, as she has done in her roles as a scientist and dean.

Robert Goldberg of the University of California at Los Angeles recalled the success of dinners before the release of reports from the National Academy of Sciences. In his view, those dinners provided a setting for nearly 100 stakeholders to gather, to hear about a report from its authors, and to have full discussions about it. He found such dinners to be respectful and ultimately more useful than speeches and press releases that occur after the release of a report. He thought that the dinners, although expensive and labor-intensive, promoted civil discourse in later public conversations, and he suggested that such dinners be reinstated. Jahn concurred that such social gatherings help diverse stakeholders to find common ground before starting conversations that may be contentious. She added that beginning with common ground does not make everyone in the room agree or care about an issue, but that "the quality of the fight is better when we can acknowledge some shared commitments before we begin the really difficult conversation," she said.

BOX 5-3
Breakout Discussion of the American Chestnut Tree

The breakout discussion report was given by Todd Kuiken, of the Woodrow Wilson International Center for Scholars.

Science Context
The American chestnut tree "was basically decimated by a blight" introduced into the United States when the Chinese chestnut was brought from China. Two competing strategies being developed to save the American chestnut so that it can be reintroduced into the wild are traditional cross-breeding with the Chinese chestnut, which has resistance to the blight, and "taking a gene from wheat, which has resistance to the blight, and inserting it into the American chestnut".

Societal Context
Most Americans probably do not know that the American chestnut has all but disappeared. A small group of American Indians is concerned about reintroduction of the chestnut, but it is unclear how much the other members of the public is engaged in or cares about this issue.

Reflections on Public Engagement
There discussed the many different frames around this topic, including the ecological implications about reintroducing a species, a positive food frame about creating a new food supply, a negative GMO frame that affects most discussions about biotechnology, and a species-conservation frame. "We may believe that scientists have successfully framed the issue in terms of what they are looking for. . . . Now, the question may be how," Kuiken said. Once this is known, the other frames come into place. In looking at all these frames, regulation and the decision-making process to create a trust factor for when these types of technologies are approved are important.

Box 5-4
Breakout Discussion of Genetically Modified Mosquitoes

The breakout discussion report was given by Trevor Butterworth of Sense about Science.

Science Context
The object of the breakout discussion was the introduction of genetically modified mosquitoes to suppress the spread of Dengue fever, Butterworth explained. Dengue is an infectious disease endemic in the developing world which, due to a warming climate, has the potential to spread along with other tropical disease into North America.

Societal Context
Oxitec, the biotechnology company that developed the genetically modified mosquito, encountered resistance in the Florida Keys about its potential release. People have concerns about the potential environmental effects of the release and the lack of communication from federal agencies conducting environmental risk assessments of a potential release. In addition, "Oxitec is a company with interests in both disease and agriculture and could be perceived as throwing a Trojan horse into the agriculture debate through the introduction of the mosquito,." Butterworth said.

Reflections on Public Engagement
Oxitec conducted public surveys to inform its engagement practices, but they seemed to have poor methods. The breakout group discussed best practices that should accompany discussions about the release of genetic-modification technologies, including "know your audience," "do not presume to know how people feel," "do not underestimate the value of consent" or the amount of time required for public engagement processes, and "use existing research in social science" and seek out the experts. Butterworth pointed out that some members of the group also thought that a "repository for best and worst practices in science communication" is badly needed.

Box 5-5
Breakout Discussion of Transgenic Corn and the Monarch Butterfly

Allison Snow, of Ohio State University, summarized the breakout discussion of transgenic corn and the monarch butterfly.

Science Context
Three waves of publications built the science story about the relationship between transgenic corn and the monarch butterfly. In 1999, John Losey, of Cornell University, published results in *Nature* of a small-scale laboratory experiment that found that one type of transgenic corn, which is not widely grown, harms monarch larvae.[a] In the same year, Tony Shelton, also of Cornell, published an article criticizing the Losey findings. The second wave occurred in 2001, when the results of six studies on a type of pollen that is widely used were published in the *Proceedings of the National Academy of Sciences of the United States of America (PNAS)*. The studies showed a lack of risk to the butterflies. The third wave is occurring today, and it involves the fear that herbicide-resistant GMOs are resulting in a decrease in milkweed, the food source for monarch caterpillars.

Societal Context
The Washington Post published an article about the Losey paper titled "Gene-Altered Corn May Kill Monarchs," and referring to the butterfly as "the Bambi of the insect world."[b] "This is a framing of unexpected effects of genetically engineered crops," Snow said. Cornell University released press statements about both the Losey and Shelton articles, but a mass-media flurry around the Losey article had already taken hold and had a dramatic adverse effect on public perceptions of transgenic corn. The six *PNAS* papers in 2001 did not receive as much publicity as the Losey paper, probably because they were published around the same time as the 9/11 terrorist attack in New York City.

Reflections on Public Engagement:
The breakout group looked across the three waves of publications and brainstormed about what could have been different. The group discussed whether scientists should be more careful about going too quickly to publication with small scale findings and, perhaps wait for results from larger scale and more thorough studies. The group also discussed whether scientists would benefit from mass-media training, especially before publishing research that could be publicly controversial. Likewise, some members of the group suggested that universities need strategic plans for times when controversial research is published. Finally the group discussed the need for science writers to consider the whole context of a story (for example, small laboratory study vs. large field study.) and take care with the frames provided by headlines. Several of the breakout group participants noted that it is hard to predict how a story will unfold, but scientists should be prepared for any possible outcome.

[a]Losey JE, Rayer LS, and ME Carter. (1999) Transgenic pollen harms monarch larvae. Nature 399, 214. doi:10.1038/20338.
[b]Weiss, R. 1999. "Gene-Altered Corn May Kill Monarchs," *Washington Post*, 20 May 1999, Page A3.

Appendix A

Workshop Agenda

DAY 1 (Thursday, January 15, 9:00am – 5:30)

SESSION 1 The Sciences of Engagement, Decisions, and Politics
Moderator: Dietram Scheufele, *University of Wisconsin-Madison*

9:00 The Science of Science Communication— Dietram Scheufele, *University of Wisconsin-Madison*

9:35 How Consumers Make Decisions — William Hallman, *Rutgers University*

10:05 Mingle and Discuss (with coffee)

10:25 Could GMOs Harm the Science Communication Environment? Vice Versa? — Dan Kahan, *Yale University*

10:55 The Role of Scientists in Policy and Politics — Roger Pielke Jr., *University of Colorado*

11:30 - 12:30 Break for Lunch

SESSION 2 Science and Perceptions: Knowns, Unknowns, and Challenges
Moderator: David Goldston, Natural Resource Defense Council (NRDC)

12:30 Public Perceptions of GM Technologies and Why it Matters –
Dominque Brossard, *University of Wisconsin* – Madison & Stephen Palacios, *Added Value Cheskin*

1:35 Science and Journalism: The Elephant in the Room — Tamar Haspel, *Food and Science Journalist*

2:05 Engaging Publics in Science and Technology — Jason Delborne, *North Carolina State University*

2:35 Mingle and Discuss (with coffee)

2:55 Panel Discussion on Science Communication and Initiatives to Label GM Foods

Moderator: David Goldston, *Natural Resource Defense Council (NRDC)*

- *Opening Comments* - David Goldston, *NRDC*
- *Panel Discussion*
  Robert Goldberg, *University of California at Los Angeles*

William Hallman, *Rutgers University*
Tamar Haspel, *Food and Science Journalist*
Eric Sachs, *Monsanto*
Allison Snow, *Ohio State University*

- *Facilitated Audience Dialogue*

4:30 Adjourn Day 1

DAY 2 (Friday, January 16, 9:00 – 2:30pm)

SESSION 3 Towards Improving the Interfaces between Scientists and Citizens

9:00 Welcome and Starting Thoughts – Brooke Smith, *COMPASS*

9:30- 11:30 Break-Out Group Discussions

Breakout A: Transgenic Corn and the Monarch Butterfly (Room 250)
Facilitator: Jason Delborne, *NC State*
Case Presenter: Jennifer Baltzegar, *NC State*

Breakout B: The American Chestnut (Room 280)
Facilitator: Sarah Evanega, *Cornell University*
Case Presenter: Rebecca Harrison, *Rennselaer Polytechnic Institute*

Breakout C: GM Mosquitoes (Members Room)
Facilitator: Trevor Butterworth, *Sense about Science*
Case Presenter: Sophia Webster, *NC State*

11:30 - 12:30 Break for Lunch

12:30 - 2:30 Reports, Reactions, Reflections
Moderator: Brooke Smith, *COMPASS*

- Report-outs
- Reaction Panel

  Rick Borchelt, *Department of Energy*
  Helene Dillard, *University of California at Davis*
  Molly Jahn, *University of Wisconsin-Madison*
  Dan Kahan, *Yale University*

- Facilitated Audience Discussion

- What Now? Conceptual and Practical Take-homes

2:30 Adjourn Workshop

Appendix B

Workshop Attendees

Ethan Alpern, Department of Energy
Richard Amasino, University of Wisconsin
Julio Araujo, unaffiliated
Jennifer Baltzegar, North Carolina State University
Juliane Baron, AERA
Cynthia Beall, Case Western Reserve University
Robin Bisson, Genetic Expert News Service
Jack Bobo, Department of State
Rick Borchelt, Department of Energy
Benjamin Boroughs, North American Miller's Association
Philip Brasher, Agri-Pulse
Evan Bromfield, Center For Food Safety
Dominque Brossard, University of Wisconsin–Madison
Amanda Buchanan, University of Maryland
Trevor Butterworth, Sense about Science
Travis Coberly, US Department of State
James Cooper, independent writer
Chris Creese, Oxitec
Jason Delborne, North Carolina State University
Helene Dillard, University of California at Davis
Sara Evanega, Cornell University
Jose Falck-Zepeda, IFPRI
Richard George, US Department of Agriculture
Robert Goldberg, University of California Los Angeles
David Goldston, Natural Resources Defense Council
Sarah Gonzalez, Agri-Pulse
Fred Gould, North Carolina State University
Ryan Green, House Science, Space, and Technology Committee
Tim Griffin, Tufts University
Annie Gutsche, DuPont Pioneer
William Hallman, Rutgers University
Jaydee Hanson, International Center for Technology Assessment
Rebecca Harrison, Rennselaer Polytechnic Institute
Tamar Haspel, food and health journalist
Keira Havens, Revolution Bioengineering
Molly Jahn, University of Wisconsin–Madison
Erik Jahner, University of California Riverside
Daniel Jenkins, Monsanto
Dan Kahan, Yale University
Dmitry Kaledin, Russian Embassy
Joe Kelsay, Dow AgroSciences
Kevin Klatt, Cornell University
John Kotcher, George Mason University
Michael Kotewicz, FDA
Todd Kuiken, Woodrow Wilson Center
Lucyna Kurtyka, Monsanto Company
David Lambert, Lambert Associates
Bruce Lewenstein, Cornell University
Michael Lohuis, Monsanto
Tiffany Lohwater, American Association for the Advancement of Science
Brian Lovett, University of Maryland
Daniel Magraw, Johns Hopkins University
Mala Mahmood, US House of Representatives
Zane Martin, National Academy of Sciences
George Matsumoto, Monterey Bay Aquarium Research Institute
Pete Matz, OFW Law
Autumn Meade, Ecological Society of America
Margaret Mellon, Mellon Associates
John Mischler, King's College
Sally Mouakkad, Research Councils UK/British Embassy
Brett Nadrich, Industry member
Clint Nesbitt, Biotechnology Industry Organization
Todd Newman American University

Ivan Oranksy, MedPage Today
Stephen Palacios, Added Value Cheskin
Molly Pfaffenroth, student
Roger Pielke Jr., Colorado State University
Kevin Pixley, CIMMYT-International Maize & Wheat Improvement Center
Suzanne Price, American Society for Nutrition
William Provine, DuPont
Keith Redin, Monsanto Company
Genna Reed, Food & Water Watch
Gary Rudgers, industry
Eric Sachs, Monsanto Company
Dietram Scheufele, University of Wisconsin-Madison
Tim Schwab, Food & Water Watch
Eden Shiferaw, OFW Law
Allison Snow, Ohio State University
Brooke Smith, COMPASS
Heven Sze, University of Maryland
Wiebke Tapken, University of Maryland
Grace Troxel, Center for Advancement of Informal Science Education (CAISE)
Michael Tu, US Department of Commerce
Melissa Varga, Union of Concerned Scientists
Sophia Webster, North Carolina State University
Ted Wells, STEMconnector
JoAnna Wendel, freelance writer
Robert Whitaker, Produce Marketing Assoication
Joe Witte, Adnet/NASA
Shunyuan Xiao, University of Maryland

Appendix C

Biographies of Workshop Speakers, Panelists, Moderators, Case Presenters, and Planning Committee Members

Jennifer Baltzegar received a M.S. in Marine Biology from the College of Charleston and is currently a Ph.D. student in the Genetics Program at North Carolina State University. She is also a NSF IGERT Fellow in Genetic Engineering and Society: The Case for Transgenic Pests where her research examines the possibility of using genetic engineering techniques that will control agricultural insect pest populations. She is particularly interested in finding effective ways to control stored grain pests.

Rick Borchelt is Director of Communications and Public Affairs at the US Department of Energy's Office of Science. Prior to DOE, he served as the Special Assistant for Public Affairs to the director of the National Cancer Institute at NIH and director of NCI's news office, providing strategic guidance and coordination of the Institute's communications and public affairs programs. Mr. Borchelt is also the former communications director for the research, education, and economics missions area of USDA, and for the USDA Office of the Chief Scientist. Prior to the USDA, he was director of communications for the Pew-funded Genetics and Public Policy Center at The Johns Hopkins University, where his work included message development, media relations, and strategic communications. He also is Lecturer in science policy and politics in the Hopkins Advanced Academic Programs division. He has had a varied career in science communications and science public policy, including stints as media relations director for the National Academy of Sciences; press secretary for the U.S. House of Representatives Committee on Science, Space and Technology under the chairmanship of the late Rep. George E. Brown, Jr.; special assistant for public affairs in the Executive Office of The President during the Clinton Administration; director of communications for the Department of Energy's Office of Science; and director of communications and public affairs at The Whitehead Institute for Biomedical Research at MIT. He is an advisor to the NSF-funded Nanoscale Informal Science Education (NISE) project, and was a committee member on the National Academy of Engineering's study of public communication about engineering. An undergraduate biology major, he's done graduate work in both insect systematics and science communication. Areas of particular interest include developing community based public engagement in science and adapting the Southern narrative tradition to science communication.

Dominique Brossard is Professor and Chair in the Department of Life Sciences Communication at the University of Wisconsin-Madison. She is on the Steering Committee of the UW-Madison Robert & Jean Holtz Center for Science and Technology Studies, and an affiliate of the UW-Madison Center for Global Studies. She is also the leader of the Societal Implications of Nanotechnology group in the National Science Foundation (NSF)-funded Nanoscale Science and Engineering Center (NSEC). Her teaching responsibilities include courses in strategic communication theory and research, with a focus on science and risk communication. Dr. Brossard's research program concentrates on the intersection between science, media, and policy. A fellow of the American Association for the Advancement of Science and a Board member of the International Network of Public Communication of Science and Technology, Brossard is an internationally known expert in public opinion dynamics related to controversial scientific issues. She is particularly interested in understanding the role of values in shaping public attitudes, and in cross-cultural analysis to understand these processes. Her lab's recent work has focused on scientific discourse in online environments, such as Twitter. She has published numerous research articles in outlets such as Science, Science Communication, the International Journal of Public Opinion, Public Understanding of Science and Communication Research. Dr. Brossard has a varied professional background including experience in the lab and in the corporate world. Notably, she spent five years at Accenture in its Change Management Services Division. She was also the communication coordinator for the Agricultural Biotechnology Support Project II (ABSPII), a position that combined public relations with marketing communication and strategic communication. Dr. Brossard earned her M.S. in plant biotechnology from the Ecole Nationale d'Agronomie de Toulouse and her M.P.S and Ph.D. in communication from Cornell University.

Trevor Butterworth is Director of Sense About Science USA, which advocates for public engagement in an evidence based approach to science and technology and for clinical trial transparency. He is also editor of STATS.org, a collaboration between the American Statistical Association and Sense About Science USA that works to improve statistical literacy in the news media. He's written for the New Yorker online, Harvard Business Review, The Financial Times, The Wall Street Journal, and many other publications. He speaks regularly about the media's coverage of science and statistics and scientific communication. He was educated at Trinity College Dublin, Georgetown, and Columbia University, and is currently a visiting fellow at Cornell University.

Jason Delborne joined North Carolina State University in August 2013 in the Chancellor's Faculty Excellence Program cluster in Genetic Engineering and Society. He serves as Associate Professor of Science, Policy and Society in the Department of Forestry and Environmental Resources and will also teach and advise students in the graduate minor program in Genetic Engineering and Society. Dr. Delborne's research focuses on highly politicized scientific controversies, such as agricultural biotechnology, nanotechnology, biofuels, and climate change. Drawing upon the highly interdisciplinary field of Science, Technology, and Society (STS), he engages various qualitative research methodologies to ask questions about how policy-makers and members of the public interface with controversial science and technology. Dr. Delborne has published peer-reviewed articles in journals such as Social Studies of Science, Public Understanding of Science, and Science and Public Policy, and he co-edited Controversies in Science and Technology: From Evolution to Energy (Mary Ann Liebert, 2010). In 2010, he received the David Edge Prize, awarded annually by the Society for Social Studies of Science (4S) for the best journal article published in the area of science and technology studies. His current project compares multiple pathways of development of genetically modified trees by exploring the extent to which "responsible innovation" is pursued and achieved. *Member of the Roundtable on Public Interfaces of the Life Sciences

Helene R. Dillard was appointed dean of the College of Agricultural and Environmental Sciences at UC Davis in January 2014. Her goal is to build upon the strengths of the college in research, teaching, extension and outreach, and maintain strong relationships with the broad range of stakeholders in California, nationally, and globally. In addition to her responsibilities as dean, she has programmatic responsibilities for the college's Agricultural Experiment Station and Cooperative Extension. Dr. Dillard has national and international leadership experience, including invited consultations, presentations, and scientific exchanges in China, Central America (Honduras, Nicaragua), South America (Argentina, Brazil, Chile), the European Union (the Netherlands, Sweden, United Kingdom), and Zimbabwe. She has worked extensively with U.S. Department of Agriculture programs, the National Institute of Food and Agriculture (NIFA), and the National Research Initiative. Prior to her appointment at UC Davis, Dr. Dillard was on the faculty at Cornell University since 1984 as a plant pathology professor, carrying a

50 percent research and 50 percent extension assignment. Her research focused on the biology, ecology, and management of fungal pathogens that cause diseases in vegetable crops. Her interests include sustainable disease management strategies, integrated pest management, epidemiology and host/pathogen/environment interactions. Dr. Dillard served as chair of the Department of Plant Pathology in Geneva, N.Y. (1997–2001), associate director of Cornell Cooperative Extension (2001–2002), and director of Cornell Cooperative Extension (2002–January 2014). She also served simultaneously as associate dean in two colleges, the College of Agriculture and Life Sciences (CALS) and the College of Human Ecology (2002–January 2014). Dr. Dillard was recognized for her contributions in plant pathology by the American Phytopathological Society (APS), receiving the Excellence in Extension Award in 1992 and being named an APS fellow in 2006. She received the New York Farmers Medal and the Outstanding Faculty Award from CALS in 2013. She completed her B.S. degree in biology of natural resources at UC Berkeley, an M.S. degree in soil science at UC Davis, and a Ph.D. degree in plant pathology at UC Davis.

Sarah Evanega received her PhD in the field of Plant Biology from Cornell University in 2009, for which she conducted an interdisciplinary study combining work in plant molecular biology with science communication. Her dissertation focused on the controversy over genetically engineered papaya in developing countries with a specific focus on Thailand. She came to Cornell after completing a BA in Biology at Reed College. Lured by great weather, plenty of water, and an unbeatable intellectual environment, she remained at Cornell University after completing her PhD to help lead a global project to help protect the world's wheat from wheat stem rust. Dr. Evanega now serves as the Director for the Cornell Alliance for Science—a global communications effort that promotes evidence-based decision-making in agriculture. She teaches courses on agricultural biotechnology at the graduate and undergraduate level. In addition, she serves as Senior Associate Director of International Programs in the College of Agriculture and Life Sciences and holds an adjunct appointment in the Section of Plant Breeding & Genetics in the Integrated School of Plant Sciences at Cornell. Sarah was instrumental in launching the CALS initiative, AWARE (Advancing Women in Agriculture through Research and Education) which promotes women in agriculture. Sarah grew up in a small agricultural village in northwest Illinois. Rebecca Harrison received a B.S. in Animal Science from Cornell University in 2014, and is now a first-year PhD student in Science and Technology Studies at Rensselaer Polytechnic Institute. She is also on the staff of the Cornell Alliance for Science. She is particularly interested in how consumers, producers, scientists, and policy-makers communicate knowledge about and respond to risk in agricultural biotechnology use. Her (1) exposure to biotechnological development at Cornell, (2) involvement with science and technology policy as a former intern with the White House Office of Science and Technology Policy, (3) engagement with the agricultural biotechnology community on the Web (independently, formerly with the Genetic Literacy Project, and currently with the Cornell Alliance for Science), (4) on-farm agricultural experience, and (5) new perspectives from the STS community have given her the standpoint necessary to appreciate not only the need for this technology, but also the need for re-envisioning how its use is communicated, and its risk regulated.

Robert Goldberg is a professor in the Department of Molecular, Cell, and Developmental Biology at the University of California Los Angeles (UCLA). He received a BS in botany from Ohio University and MS and PhD degrees in plant genetics from the University of Arizona. He was a postdoctoral fellow at the California Institute of Technology. Dr. Goldberg has received several awards at UCLA: the Distinguished Faculty Teaching Award from the Biology Department, the Distinguished Faculty Teaching Award from the Division of Molecular and Cell Biology, the Luckmann Distinguished Teaching Award from the Academic Senate and Alumni Foundation, and the Gold Shield Faculty Research Award for Excellence in Undergraduate Education and Research. In addition, he was honored to have been named one of the top 20 professors in UCLA' s 75-year history. He was also awarded the National Order of Scientific Merit Grà Cruz from the President of Brazil. In 2001, he was elected to membership in the National Academy of Sciences. Dr. Goldberg created The Plant Cell, organized the first plantoriented Keystone Meetings, and served as program director of several U.S. Department of

Agriculture Plant Genetics and Crop Improvement Grant panels. More recently, he established the Seed Institute, a multiuniversity collaboration dedicated to "uncovering all the genes required to make a seed" which is the focal point of his present research efforts. He has also been the director of the American Society of Plant Biologists Education Foundation.

David Goldston is Director of Government Affairs for Natural Resources Defense Council in Washington, D.C. and is responsible for its governmental strategies, bringing together NRDC's interactions with Congress, the administration and the public. He has more than twenty years of experience on Capitol Hill, working mainly on science and environmental policy and served as chief of staff of the House Committee on Science from 2001 through 2006. He has been a visiting lecturer at Princeton and Harvard Universities and a columnist for the journal Nature. In 2008 and 2009, he was project director for the Bipartisan Policy Center report, "Improving the Use of Science in Regulatory Policy" and he has served on several panels at the National Academy of Sciences. David graduated from Cornell University in 1978 with a B.A. in history and completed the course work for a Ph.D. in American history at the University of Pennsylvania.

William K. Hallman is a professor and Chair of the Department of Human Ecology and is a member of the graduate faculty of the Department of Nutritional Sciences, and of the Bloustein School of Planning and Public Policy at Rutgers, the State University of New Jersey. He is a 1983 graduate of Juniata College in Huntingdon, Pennsylvania and earned his PhD. in Experimental Psychology from the University of South Carolina in 1989. Dr. Hallman's research examines public perceptions of controversial issues concerning food, health, and the environment. Recent research projects have looked at consumer perceptions and behaviors concerning genetically modified foods, animal cloning, avian influenza, accidental and intentional food contamination incidents, and food recalls. His current research projects include studies of public perceptions and responses to food safety risks, the safety of fresh meat, poultry, game, and seafood products purchased on the Internet, the use of nanotechnology in food, and public understanding of health claims made for food products. Dr. Hallman serves on the Executive Committee of Rutgers Against Hunger (RAH), and helped to found the New Brunswick Community Farmers Market, which offers food insecure residents access to fresh, locally grown, affordable, nutritious, and culturally appropriate produce and other food products. Dr. Hallman formerly served as the Director of the Food Policy Institute (FPI) at Rutgers, and currently serves as the Chair of the Risk Communication Advisory Committee of the US Food and Drug Administration (FDA

Tamar Haspel is a journalist who's been on the food and science beat for the best part of two decades. She writes a monthly Washington Post column, Unearthed, which covers food supply issues: biotech, pesticides, food additives, antibiotics, organics, nutrition, and food policy. When she's tired of the heavy lifting of journalism, Ms. Haspel helps her husband on their oyster farm, Barnstable Oyster.

Molly Jahn is a Professor in the Laboratory of Genetics and Department of Agronomy at the University of Wisconsin-Madison, and Special Advisor to the Chancellor and Provost for Sustainability Sciences. She has had a distinguished research career in plant genetics, genomics and plant breeding of vegetable crops focusing on molecular genetics of disease resistance and quality traits. Her research groups at UW Madison and Cornell University have produced crop varieties now grown commercially and for subsistence on six continents under approximately 60 active commercial licenses. She has also worked extensively in developing countries to link crop breeding with improved human nutrition and welfare. Her innovative approaches to inter-sector partnerships, engagement with emerging institutions, and integrated projects focused on impact and technology transfer have been highlighted in numerous studies and books. She has consulted widely in the private sector, and has served as an advisor for philanthropic interests, venture capital and finance, First Nations, and U.S and foreign governmental agencies in agriculture, food security, life and environmental sciences. She received the BA with distinction in biology from Swarthmore College and holds graduate degrees from MIT and Cornell University.

Dan Kahan is the Elizabeth K. Dollard Professor of Law and Professor of Psychology at Yale Law

School. He is a member of the Cultural Cognition Project, an interdisciplinary team of scholars who use empirical methods to examine the impact of group values on perceptions of risk and science communication. In studies funded by the National Science Foundation, Professor Kahan and his collaborators have investigated public dissensus over climate change, public perceptions of scientific consensus across disputed issues, and public reactions to emerging technologies. Articles featuring the Project's studies have appeared in a variety of peer-reviewed scholarly journals including the Journal of Risk Research, Judgment and Decision Making, Nature Climate Change, Science, and Nature. The Project is currently engaged in a field research that features using evidencebased strategies to promote public engagement with climate science in Southeast Florida.

Stephen Palacios is an Executive Vice President with the innovation consulting firm, Added Value Cheskin. He leads the company's Hispanic practice, directing strategy on client engagements relating to new market assessment, product innovation, and communication strategy. Clients include Pepsi, Wells Fargo, Time Warner Inc., AstraZeneca. He is a leading expert in the U.S. Hispanic market having helped guide strategy for organizations such as Blue Cross Blue Shield (various regions) Meredith Corporation and the National Council of La Raza. Mr. Palacios holds a B.A. from Saint Joseph's University (PA), where he was Valedictorian and an M.A. from American University, where he was awarded a Fellowship. He is a frequent speaker at industry conferences and has been featured in publications including the Los Angeles Times, Harvard Business Review and AdAge, and has been featured on ABC's Nightline and PBS's Latino market documentary, Brown is the New Green.

Roger Pielke Jr. has been on the faculty of the University of Colorado since 2001. He is a Professor in the Environmental Studies Program and a Fellow of the Cooperative Institute for Research in Environmental Sciences (CIRES). Dr. Pielke's research focuses on science, innovation and politics. In 2011 began to write and research on the governance of sports organizations, including FIFA and the NCAA. He holds degrees in mathematics, public policy and political science, all from the University of Colorado. In 2012 Dr. Pielke was awarded an honorary doctorate from Linköping University in Sweden and was also awarded the Public Service Award of the Geological Society of America. He also received the Eduard Brückner Prize in Munich, Germany in 2006 for outstanding achievement in interdisciplinary climate research. Before joining the faculty of the University of Colorado, from 1993-2001 Dr. Pielke was a Scientist at the National Center for Atmospheric Research. He is also author, co-author or co-editor of seven books, including The Honest Broker: Making Sense of Science in Policy and Politics published by Cambridge University Press (2007) and The Climate Fix: What Scientists and Politicians Won't Tell you About Global Warming (2010, Basic Books). His most recent book is Rightful Place of Science Series, Disasters and Climate Change (2014, Consortium for Science, Policy & Outcomes). He is currently working on a book on sports in society.

Eric S. Sachs earned a PhD in Genetics at Texas A&M University, and MS and BS degrees in Botany from the University of California, Davis. He has worked at Monsanto Company, St. Louis, for 36 years and has played key roles in the development, authorization and commercial application of GM crops. He currently focuses on the Social, Economic and Environmental impacts of GM cropping systems, as well as supporting science-based regulatory systems and Monsanto's sustainability strategy. He is responsible for communicating science-based environmental risk assessment, environmental safety, and social and economic impacts of Monsanto technologies to stakeholders through outreach and partnership with external individuals and organizations, which includes expanding, equipping and empowering experts to communicate effectively about agricultural systems and GMOs to key stakeholder groups and the public. As a leader and communicator within the private sector, he successfully uses his knowledge of science and biotechnology, experience and passion to communicate the safety and benefits of GM crops, to demystify the science of biotechnology and to build confidence among policy makers, opinion leaders and the public.

Dietram Scheufele is the John E. Ross Professor in Science Communication in the Department of Life Sciences Communication at the University of Wisconsin, Madison, and Co-PI of the Center for Nanotechnology in Society at Arizona State University. His research focuses on shaping public attitudes toward science and technology, with emphasis on the role that social media and other emerging modes of communication play in society. Dr. Scheufele has published extensively in the areas of public opinion, political communication, and public attitudes towards emerging technologies, including nanotechnology, synthetic biology, stem cell research, nuclear energy, and genetically modified organizes. Dr. Scheufele has served on many committees and advisory panels, including the National Conference of Lawyers and Scientists, the Nanotechnology Technical Advisory Group to the U.S. President's Council of Advisors on Science and Technology, and the National Academy of Engineering Committee on Developing Effective Messages for Improving Public Understanding of Engineering. He is also a co-chair of the National Academy of Sciences, Roundtable on Public Interfaces of the Life Sciences. Dr. Scheufele received both his MA Journalism and Mass Communications and his PhD in Mass Communications from the University of Wisconsin-Madison.

Brooke Smith is the Executive Director of COMPASS, a science communications organization focused on helping scientists be more effective communicators, and helping scientists engage with society and the public discourse about the environment. Ms. Smith's career has focused on being a practitioner of science communications, a sustainability leader, and a nonprofit executive. Her experiences are in ocean and environmental science, state and federal environmental policy, environmental consulting, communications, connecting science to policy and management, and nonprofit leadership and management. Ms. Smith leads COMPASS in vision, strategy, fundraising and administration. She received her MS from Oregon State University's College of Oceanic and Atmospheric Sciences, and her bachelor's degree from Duke University. She holds a courtesy faculty appointment at Oregon State University, serves on the Board of Directors of Portland's locally based Forest Park Conservancy, recently served on the National Board of Directors of the Surfrider Foundation and was recently a Donella Meadows Leadership Fellow. She lives in Portland OR with her husband and their 2 daughters.

Allison Snow is professor of evolution, ecology and organismal biology at The Ohio State University. Her Plant Population Ecology Lab studies natural selection and ecological processes within plant populations, including the dynamics of gene flow, especially involving transgenic plants. Trained as a plant ecologist at the University of Massachusetts, Dr. Snow received postdoctoral fellowships from the National Science Foundation and the Smithsonian Institution. Her current research combines molecular and ecological approaches to understand how quickly crop genes move into wild populations, and the extent to which novel transgenic traits could benefit weedy and semi-weedy plants. She is the lead author of a 2005 Ecological Society of America position paper on environmental effects of genetically engineered organisms. A fellow of the American Association for the Advancement of Science and the Aldo Leopold Leadership Program, she has served on the editorial boards of Ecology, Ecological Monographs, Evolution and Environmental Biosafety Research. A past president of the Botanical Society of America, she has served on the U.S. Department of Agriculture's National Genetic Resources Advisory Board and panels convened to discuss issues in transgenic organisms by the National Research Council and the Academy of Finland. In 2002, she was one of Scientific American's Top 50 Researchers in Science and Technology. She also directs the Undergraduate Research Office at Ohio State.

Sophia Webster, originally from Arlington, Virginia, received her B.S. in Biology with minors in Chemistry and Entomology from Virginia Tech in 2012. Sophia is part of the first cohort of students at North Carolina State University supported by the NSF IGERT training grant on Genetic Engineering and Society: The Case of Transgenic Pests. Her research is on (1) Development of killer-rescue gene drive systems in the dengue fever vector Aedes aegypti and (2) Evaluation of the reduce & replace model in Drosophila melanogaster.

Appendix D

About the Roundtable on Public Interfaces of the Life Sciences

The Roundtable on the Public Interfaces of the Life Sciences (PILS) of the National Research Council is a forum that seeks to monitor and improve understanding of the intersections between different life science communities and public audiences on topics that spark public concerns, generate policy debates, or influence market dynamics. The overarching vision of the roundtable is for life scientists to understand the dynamics of public interfaces, and have access to the knowledge and tools needed to develop proactive, collaborative, science-based approaches to public interfaces about emerging topics in the life sciences.

The PILS Roundtable is an active and engaged network that brings together research life scientists, social scientists studying science communication, and professional science communicators. It provides leadership to the life science community through activities that raise awareness among life scientists about the importance of public interfaces; encourage networks among life scientists, communication scientists, informal education experts, and science communicators; and facilitate the development of partnerships and other initiatives among PILS members and their institutions to improve public interfaces for current and emerging life sciences issues.

The PILS Roundtable is led by the National Research Council's Division on Earth and Life Studies in partnership with the Division on Behavioral and Social Science and Education.